ENVIRONMENTAL EFFECTS

OF

SURFACE MINERAL WORKINGS

Report to:

Department of the Environment

Prepared by:

Roy Waller Associates Ltd

UNIVERSITY OF PLYMOUTH
DEPARTMENT OF GEOLOGICAL
SCIENCES
Copy Number: 2

October 1991

London: HMSO

© Crown copyright 1992
Applications for reproduction should be made to HMSO
First published 1992

ISBN 0 11 752637 1

FOREWORD

I welcome this report.

Minerals are an essential commodity, but working them can be disruptive and a nuisance to local residents. This report provides a valuable guide to the issues that need to be considered by local authorities and operators.

Of course not all the suggestions will be applicable to all situations. Nevertheless, I hope the report will provide a useful stimulus to both industry and local authorities to seek ways of securing better environmental standards.

Tim Yeo
Parliamentary Undersecretary of State for the Environment
March 1992

EXECUTIVE SUMMARY

The Department of the Environment commissioned this study to review the environmental effects which occur during the surface winning of minerals. The study was based primarily on a literature review and complemented by a series of site visits and discussions. The objectives were to identify the information describing good practice and to bring it to the attention of mineral operators, planning authorities and communities affected by the surface minerals industry. Where the best of existing practice or understanding is not good enough to avoid significantly adverse effects, the study suggests topics for research and Departmental advice.

Apart from limited exceptions, ie blasting, noise and to some extent dust and landscape, the literature on environmental effects and the related good practice which can be used to ameliorate them is not extensive.

A number of general issues do however come out quite strongly as needing more attention, namely enforcement, incremental effects, compensatory measures and the need to plan to avoid adverse effects. These issues are only indirectly within the terms of reference of this study. Nonetheless, all have a major effect on the level of response, in the form of complaints, occasioned by environmental problems.

Many problems and adverse effects can be mitigated by integrating environmental considerations into the overall planning and by the development of balanced, well-worded and enforceable planning conditions.

Unfortunately, there is sometimes insufficient information about potential effects and not enough monitoring to give warning of impending problems for preventative action to be taken. There are also several difficulties associated with **enforcement**:
- compliance with and the enforcement of planning conditions seem to receive less attention than their consideration at the planning application stage,
- enforcement procedures are time consuming and may have uncertain results,

Environmental Effects of Surface Mineral Workings

Summary

- there is concern that conditions, or indeed Section 106 agreements, should not duplicate statute law and the duties of other bodies, eg the Police, HM Inspectors of Quarries; this sometimes causes issues to fall between two stools with no party taking the initiative.

Operator self-regulation is often relied upon, eg to ensure that vehicles entering or leaving a site do so on identified routes, or are sheeted to prevent dust and spillage. Unfortunately not all operators are sufficiently interested that they volunteer or accept the idea of self-regulation.

Operators should be encouraged to undertake a degree of **self-regulation** and to carry out or to commission periodic environmental reviews or audits. These could cover compliance with any conditions of the Consent, response to complaints, a description of existing site practice and any ameliorative measures taken or proposed. The results should preferably be published. This approach would facilitate, but not replace, the work of the enforcement agencies and release resources to be used where they are most needed. It would also mean that audits would form part of an operator's quality assurance regime and would improve the anticipation and avoidance of difficulties.

The **duration** of noise, dust, traffic, blasting, etc to which residents and others are subject is clearly important. Whilst Minerals Local Plans include policies regarding the totality of workings in an area, there is no framework which allows for consideration of the **incremental effects** of a series of operations, taking place one after the other, in the vicinity of a single community. A series of workings, each of which may last a few years, can result in decades of continuous exposure to traffic and visual intrusion. The total incremental effect may be worse than a single integrated and progressively restored working. Research is needed into the significance of incremental effects and ways in which it would be possible to reflect such effects in national and local policies. It would also help the planning process to have a better understanding of the concerns underlying peoples' antipathy; settling specific objections and individual complaints does not always resolve the problem.

The better use of **Minerals Local Plans** seems to be one way in which the incremental effects of a number of workings in one area might be properly managed and of providing a means whereby local people can have an input into the planning of their area. They provide an opportunity to keep sensitive development away from workings and to avoid inhibiting their progress. This is especially true of hard rock quarries because of their long life. Minerals Local Plans are also a way of defining those areas within which development will not be permitted in order to avoid sterilising mineral resources as well as defining those areas in which there will be opposition to applications for workings.

The reaction of residents to workings is affected by the quality of the **communication** between them and the operator both on a personal level and through liaison committees. It is also influenced by the image that they have of the operator, including whether they perceive the operator as wanting to be a good neighbour or just to placate them.

Compensatory measures can reduce or offset the environmental disturbance that a community suffers during the life of workings. They can alter the context in which problems are seen and may significantly reduce the level of complaint. Compensatory measures may help individual residents, eg soundproofing of their homes, or the community as a whole through the provision of some form of 'planning gain'.

The specific operational problems created by surface mineral workings based upon **public complaint** are, in approximate order of significance, traffic, blasting, noise, dust, visual effects. Other matters, ie, water, wastes, ecology, archaeology, agriculture and forestry may be equally important environmental considerations but are less usually the subject of public protest.

The potential for problems to occur varies significantly with the type of mineral working and site-specific factors. Any generalisation can only be a guide to the issues to be considered. For particular sites many issues will be satisfactorily resolved at the planning stage. In the longer term coherent mineral-specific **codes of practice** would be helpful.

Road traffic is an almost universal source of difficulty where it is the main means of transport. Alternatives to road haulage may also have difficulties, eg at the distribution end of a sea-link or from rail traffic at night. The responsibility for the control of traffic and its effects lies with the police. Highway Authorities can limit road use by vehicle size and time restrictions but these must apply to all users. Operators can and do impose conditions upon their own and sub-contractors' vehicles and drivers; some impose conditions on their clients. More cooperation between the regulatory authorities is desirable. Some way of facilitating control is necessary, eg a requirement to sheet a vehicle is easier to enforce than a requirement that there should be no spillage. Good practice by operators may include voluntary routeing arrangements for themselves and their clients, the provision of overnight on-site parking to avoid early morning arrivals and queuing, the provision of vehicle washing and sheeting facilities.

Blasting, where it occurs, can be a major concern. At the present, ground vibration is the main cause of complaint. Whilst it is true that residents, particularly owner occupiers, are likely to complain less if they are convinced that there is no risk of damage, it is clear that complaints of disturbance can still be significant. Operators

Summary

need to accept that disturbance is a justifiable cause for complaint even when there is no risk of damage. Overpressure, which may be confused with ground vibration, also causes complaint. This may become relatively more significant if levels of ground vibration are further reduced. There is a need for guidance on what levels of vibration and overpressure produce complaint and nuisance. Fly rock is often said to be unpredictable and uncontrollable. The latter is clearly untrue. It is the responsibility of HM Inspectors of Quarries and the Health and Safety Executive to ensure that quarry management complies with the Quarries (Explosives) Regulations 1988. These require blasting specifications to be produced for preventing danger from blasts. There is a need for Mineral Planning Authorities to agree reasonable distances between quarries and neighbours with the Inspectorate. In terms of good practice, there is some possibility of limiting the need to blast by the use of modern 'earthmoving' plant, but the main way of controlling vibration and overpressure will be by good blast design and practice. Surface detonating cord and plaster/secondary blasting should be avoided. Maximum instantaneous charges should be minimised, eg by decking charges, and greater 'factors of safety' used when blasting in variable conditions or unusual situations. To reduce feelings of apprehension produced by vibration, structural-condition surveys should be available to neighbouring householders prior to the commencement of blasting. These will help quickly and convincingly to show whether or not any damage has occurred and to ensure that compensation is paid if appropriate.

Noise has been the subject of a separate research study sponsored by the Department. People complain of early starts; being woken up by lorries arriving at site or machines starting-up can be particularly annoying. An alternative to accepting public pressure for later starts would be to consider a transitional dawn period, and to define acceptable noise levels and/or activities comparable to the more common evening restrictions. Operators should be more willing to take noise emission levels into account when buying or hiring plant and equipment and when choosing methods of working. An industry-wide data bank, especially for mobile plant, is desirable. Choosing quiet plant or retrofitting existing plant to make it quieter involves more than making sure that it has an effective exhaust silencer or fitting a better one, eg reversing alarms are a particular source of nuisance. Whilst many alternative forms of alarm exist, guidance on their acceptability in terms of safety and the balance of their environmental impact is needed, eg the effect of flashing lights at night. We suggest that more consideration is given to the use of temporary/permanent noise screens, particularly around individual houses, rather than the use of mounds. In some cases it may be valuable to use the two together. In extreme cases, noise can be controlled at residents' houses by the provision of secondary windows.

Dust emanating from a site is difficult to measure and is often difficult to control. Much of the control exercised by operators appears to be a matter of trial and error. Many ways of controlling dust have been described in the literature but few data on experience of their use and cost-effectiveness. Dust can be avoided by operational planning, eg by using conveyors not dump trucks, by creating 'sensitive zones' of the site within which activities will be limited, and by early vegetation and restoration to minimise exposed surfaces. It can be controlled by enclosure, collection, good housekeeping, watering and by removal from the atmosphere using water sprays and mists. In very dry and windy weather it may be necessary to stop some activities altogether. It is evident that the use of water without additives to ensure the wetting of dust and the retention of moisture, may adversely effect the quality of the mineral product. More data on the effectiveness of additives and formulations requiring less water, eg fine sprays, micro-foams, are needed. There is also a need for research into performance criteria, measurement techniques and acceptable levels of dust, together with guidance on the use of additives including any environmental side effects.

Landscape and visual intrusion are subjects which arouse considerable passion at the planning application stage but form the basis of relatively few complaints during working. Once planning permission has been granted, residents seem to turn their attention to more tangible day to day matters, such as noise and dust. Nonetheless it remains a significant issue. The planning decisions on size and sequence of working usually determine most of the loss of landscape; conditions for planting and screening limit the visual intrusion. Screening and planting can be very successful, although conditions requiring preplanting or planting are sometimes met by token planting which is ineffective and incongruous or are ignored and unenforced. Progressive restoration is a key element in minimising operational impact; it should commence as early as possible in the life of the workings and preferably be coupled with pre-planting to soften the initial impact. Site layout, landscaping, design and colour of processing plant, use of low-profile plant in sensitive situations and direction of working all need to be considered. Other factors are the location of haul-roads, areas for the disposal of waste and temporary storage of soil and overburden. At many sites the design of the entrance will be particularly important, especially if the processing plant is adjacent to it. Good housekeeping will be vital in maintaining a good appearance. A balance should be drawn between the use of man-made bunds, fences, planting and other measures to screen the site from properties. However, in some cases, such features might be as alien to the neighbourhood as the workings. More research is desirable on the optimum sequencing of mining operations to facilitate progressive restoration without significantly increasing operating costs.

Numerous potential water problems are described in the literature but the situation in practice is less clear. The extent of **groundwater** problems, ie derogation, loss of aquifer capacity, settlement, contamination, may not be great but it is likely to increase as workings become deeper. Good practice in relation to groundwater should recognise the possible effects of dewatering and of working close to underlying aquifers. In many cases, monitoring will give a clearer picture of potential problems before work starts, enabling control measures to be incorporated in the design. There is need for a code of practice regarding monitoring requirements and the precautions to be taken. Control measures include:- avoiding dewatering where possible; leaving lateral and vertical margins between the workings and sensitive neighbouring features, eg aquifers, abstractions, surface water courses, vegetation; providing impermeable sites for the containment of spillages of chemicals, oil, etc. In some cases, compensatory measures, such as alternative water supplies, should be provided.

The main problems with **surface water** are flashier flows and discharges polluted with excessive suspended solids and in some cases chemicals. There is often a lack of a scientific design of settling ponds and poor integration into the landscape. The main way to control surface water flows and their quality is to have an integrated water system, including large enough sumps to provide a buffer store of water for flow balancing by means of pumps and well-designed settling lagoons. The pick-up of solids by rainwater running off exposed soil surfaces can be reduced by protecting exposed surfaces, eg by vegetating, as soon as possible. The performance of the lagoons can be enhanced by complementary chemical treatment. There is a need for more guidance on desirable standards of discharges especially in storm conditions and a code of practice for the technical and visual design of lagoons.

Above all, for both groundwater and surface water, there is a need for more baseline monitoring to provide enough information for the prediction and avoidance of serious problems and to help understand why residual problems occur.

Waste, overburden and soils, if inappropriately placed, can be very unsightly and can adversely affect flood regimes. They can also be a source of suspended solids in run-off but this is controllable. Silt can be dangerous until dried out and covered. Good practice may be to create less waste, to use it beneficially, to include it in progressive restoration or to landscape and vegetate it.

Severance of footpaths, animal runs can occasionally create difficulties. It is best to plan to avoid problems, to provide alternatives and to maintain them.

A computerised data base of the literature reviewed during the study has been created. It is held by the Department of the Environment.

CONTENTS

1. **INTRODUCTION**
   - 1.1 Scope of Study 1
   - 1.2 Methodology 2
   - 1.3 Outline of Report 2
   - 1.4 Introductory Comments 3

2. **ENVIRONMENTAL CONCERNS & PROBLEMS**
   - 2.1 Nature of Environmental Concerns and Problems 5
   - 2.2 Management of Concerns and Control of Problems 8
   - 2.3 Related Issues 14
   - 2.4 Recommendations for Further Advice and Research 14

3. **TRAFFIC**
   - 3.1 General 17
   - 3.2 Problems 17
   - 3.3 Acceptable Levels and Monitoring 18
   - 3.4 Good Practice 18
   - 3.5 Recommendations for Further Advice and Research 24
   - *Summary of Good Practice* 25

4. **BLASTING – VIBRATION, OVERPRESSURE & FLYROCK**
   - 4.1 General 27
   - 4.2 Problems 28
   - 4.3 Acceptable Levels and Monitoring 29
   - 4.4 Good Practice 34
   - 4.5 Recommendations for Further Advice and Research 40
   - *Summary of Good Practice* 41

5. **NOISE**
   - 5.1 General 43
   - 5.2 Problems 43
   - 5.3 Acceptable Levels and Monitoring 44
   - 5.4 Good Practice 49
   - 5.5 Recommendations for Further Advice and Research 54
   - *Summary of Good Practice* 56

6. **AIR POLLUTION – DUST & ODOUR**
   - 6.1 General 57
   - 6.2 Problems 58
   - 6.3 Acceptable Levels and Monitoring 60
   - 6.4 Good Practice 62
   - 6.5 Recommendations for Further Advice and Research 68
   - *Summary of Good Practice* 69

7. **LANDSCAPE – VISUAL INTRUSION**
   - 7.1 General 71
   - 7.2 Problems 71
   - 7.3 Acceptable Measures and Monitoring 72
   - 7.4 Good Practice 73
   - 7.5 Recommendations for Further Advice and Research 77
   - *Summary of Good Practice* 78

8. **GROUNDWATER**
   - 8.1 General 79
   - 8.2 Problems 79
   - 8.3 Acceptable Changes and Monitoring 82
   - 8.4 Good Practice 83
   - 8.5 Recommendations for Further Advice and Research 86
   - *Summary of Good Practice* 88

9. SURFACE WATER
- 9.1 General ... 89
- 9.2 Problems ... 89
- 9.3 Acceptable Changes and Monitoring ... 92
- 9.4 Good Practice ... 93
- 9.5 Recommendations for Further Advice and Research ... 97
- *Summary of Good Practice* ... 98

10. WASTES
- 10.1 General ... 99
- 10.2 Problems ... 100
- 10.3 Acceptable Situations ... 101
- 10.4 Good Practice ... 101
- 10.5 Recommendations for Further Advice and Research ... 102
- *Summary of Good Practice* ... 103

11. SEVERANCE AND FOOTPATHS
- 11.1 General ... 105
- 11.2 Problems ... 105
- 11.3 Acceptable Changes and Monitoring ... 106
- 11.4 Good Practice ... 106
- 11.5 Recommendations for Further Advice and Research ... 107
- *Summary of Good Practice* ... 108

12. EFFECTS ON RESIDENTS
- 12.1 General ... 109
- 12.2 Cooperation and Community Links ... 112
- 12.3 Operator's Image ... 113
- 12.4 Compensatory Measures ... 114
- 12.5 Duration of Nuisance, Incremental Effects ... 116
- 12.6 Hours of Work ... 117
- 12.7 Buffer Zones and Local Plans ... 118
- 12.8 Recommendations for Further Advice and Research ... 120
- *Summary of Good Practice* ... 121

13. EFFECTS ON OTHER SECTORS
- 13.1 Visitors, Tourists, Investors & Heritage ... 123
- 13.2 Agriculture & Forestry ... 124
- 13.3 Ecosystems/Wildlife ... 125
- 13.4 Archaeology ... 126
- 13.5 Recommendations for Further Advice and Research ... 128
- *Summary of Good Practice* ... 129

14. DISCUSSION OF RELATED MATTERS
- 14.1 Planning Techniques ... 131
- 14.2 Trends ... 132
- 14.3 Monitoring & Enforcement ... 134
- 14.4 Recommendations for Further Advice and Research ... 138
- *Summary of Good Practice* ... 139

15. SUMMARY OF RECOMMENDATIONS FOR FURTHER ADVICE AND RESEARCH
- 15.1 General ... 141
- 15.2 Advice ... 141
- 15.3 Criteria and Standards ... 142
- 15.4 Codes of Practice ... 143
- 15.5 Other Research ... 143

16. ACKNOWLEDGEMENTS ... 145

17. GLOSSARY ... 149

18. REFERENCES (numerical) ... 153

19. REFERENCES (alphabetical) ... 165

CHAPTER 1

INTRODUCTION

1.1 SCOPE OF STUDY

This study was commissioned by the Department of the Environment to review the environmental effects which occur during the surface winning of minerals. The need for the study arose from a general increase in public awareness of environmental matters and a demand for higher standards, the increasing pressure on the provision of land for mineral extraction and the consequent difficulties that face operators in seeking planning approval.

Whilst the considerable progress of the industry and many individual operators in reducing environmental effects is acknowledged, this study is concerned with the problems that remain. These have been addressed by examining the situation that existed in the late 1980s. There has been no attempt to present an historical review.

The main objectives were to:
- provide information to the surface minerals industry and those affected by it,
- help ensure that the best use is made of existing good practice and techniques,
- suggest additional advice to be provided by the Department,
- identify key problem areas where current best practice is insufficient,
- propose further research to overcome these shortcomings.

The study covered the effects of traffic, blasting, noise, dust, wastes, severance and the effects on visual amenity, water, archaeology, agriculture and ecology. It considered the main surface worked minerals in England and Wales: coal, limestone, chalk, clay, peat, hard rock, sand and gravel. Metalliferous mining has not been considered because of its small scale and specialised nature.

Considerations of competing land use, the subsequent use of sites for waste disposal and restoration were specifically excluded. Where operational activities affect subsequent restoration, these have been mentioned only in passing. Mineral processing has been covered only where it is conventionally integrated with mineral winning and then only in general terms.

1. Introduction

In summary, this was a preliminary broad study to provide the basis for a programme of further work and to changes in advice, in operating practice and in the statutory planning framework. **Importantly this report is not a code of practice.**

1.2 METHODOLOGY

The study was based primarily upon a literature survey. Material was sought via journals, libraries and computerised data bases. Information was canvassed directly by extensive circulation, including all Mineral Planning Authorities and some other Local Authorities, operators, consultants and research institutions. A general call for information was made through the trade and technical press. In all some 2000 references have been scanned and approximately 600 have been studied in detail. Abstracts have been produced for 500 which are relevant to the issues covered in this report and their details entered in a computerised data base. In addition, 40 written responses and comments were received.

The literature survey was complemented by many meetings and a number of site visits which proved to be important because a significant part of the relevant experience is not recorded in a published or publishable form.

Where concerns or problems have been identified, the study has not been inhibited by past Departmental advice, precedents created by previous planning decisions, previously 'accepted' practice, or existing or proposed legislation.

1.3 OUTLINE OF REPORT

The environmental concerns associated with mineral workings are identified and broadly discussed in Chapter 2. These environmental effects are then discussed individually in Chapters 3–11 and the structure of these chapters is:
- a *general review* of potential effects,
- reports of *problems* met and *concerns* expressed,
- where appropriate, *acceptable levels* of pollution and the means of monitoring them,
- suggestions for *good practice*,
- *recommendations* for further advice and research,
- a *summary* of good practice.

Later, Chapters 12–14 discuss:
- the factors which affect a community's perception of, and reaction to, adverse environmental effects,
- the aggregated effects upon other impacted sectors, eg investors, agriculture, forestry, flora and fauna, archaeology and heritage,
- trends in mining practice,
- difficulties of monitoring and enforcement,
- buffer zones,
- the use of local plans.

Chapter 15 is a summary of suggestions and recommendations for further advice and research. The most frequently used abbreviations and terms are explained in Chapter 17.

Where written sources are used, they are referred to throughout. Not all the abstracted material is referred to in the text. Observations from site visits or statements made by third parties rarely lend themselves to direct referencing. Generally, such information will be referred to in the third person, eg *'It has been observed that ...'*. Where Roy Waller Associates Ltd is expressing an opinion, the first person will be used, eg, *'We believe that ...'*.

The references listed in Chapters 18 & 19 are provided in both alphabetical order of authors names and in numerical order. A computerised data base of the material listed in this report is maintained within the Department of the Environment. Details of access to the data base can be obtained by writing to:- Minerals Division, Room C15/20, 2 Marsham Street, LONDON, SW1P 3EB.

1.4 INTRODUCTORY COMMENTS

Many of the problems are common in varying degrees to all minerals, eg, those due to traffic on public highways, where this is the main means of transport. This report is therefore centred on the various environmental effects rather than on the different sectors of the minerals industry. Reference is made to specific minerals only where effects are materially different. It is important to note that working practices vary within a mineral sector as the result of the age of the working and the nature of the operator as well as between mineral sectors.

Problems sometimes arise because operations have been permitted to take place too near to houses, or vice versa, however, they often occur because the practices employed are not good. **'Good practice', in the context of this report, is a compilation of practice which various authors have suggested or reported as being useful in avoiding or reducing problems, with the explicit or implicit suggestion that the associated costs are not excessive.**

1. Introduction

Descriptions of good practice are therefore confined to those practices which are available and which may be beneficial in the amelioration of effects. No judgements have been made about the reasonableness or otherwise of the associated costs. There is no intention that the suggestions for good practice can or should be employed in every case; simply that these possibilities exist and should be considered in the context of the local situation. The degree of improvement required and the usefulness of the available good practices will vary with the type of mineral, the site and the neighbouring environment.

It is common experience that the use of new techniques when applied for purely environmental reasons is initially more expensive. However, once an operator/manufacturer has become accustomed to them, advantages and opportunities offered by the new 'tools' are often discovered. An example is the well-established practice of noise bunds which provide a means of storing material.

In our visits and meetings examples were found where better practice could have been used to advantage. Unfortunately there is little information available about the extent to which the 'good practice' identified in this report is used, or needed but not used. Many operators quite reasonably tend to use their own staff for solving _ad hoc_ problems to minimise costs and to maximise their own understanding of the issues. This and commercial interests may have inhibited the publication of useful information.

Not all the recommendations for further work are directed towards the Department. Some work should be undertaken by industry associations, British Standards Institution (BSI) or _ad hoc_ working parties, eg codes of practice and data bases. Specific technical problems could be addressed by industry associations, individual operators or manufacturers.

CHAPTER 2

ENVIRONMENTAL CONCERNS & PROBLEMS

2.1 NATURE OF ENVIRONMENTAL CONCERNS AND PROBLEMS

2.1.1 General

The adverse effects of a working can be significant even though the operations conform with a standard, are within the terms of a planning consent, or conform with a code of practice. Any standard is a compromise between the costs of control and the benefits to be gained; a standard rarely seeks to avoid any adverse impact. Planning consents are likely to take account of local conditions and sensitivities. Even so the consequential effects may be significant and important to neighbours and, as such, should be considered. Equally an existing facility, eg a use of land for recreation or wildlife, may not be a legally protected entitlement but its loss would be an adverse effect for those involved. In this report we try to avoid over emphasising legislation and standards but try instead to discuss concerns and problems in more colloquial terms.

The incidence of environmental concerns and problems associated with surface mineral workings are not comprehensively reported in the literature and examples of actual problems are often anecdotal. In some cases problems are anticipated, eg in Environmental Assessments, but the results, if any, of subsequent monitoring are not available to confirm the extent to which the predicted problems occurred or vice versa. Nonetheless, it is possible to get a useful understanding of the relative importance of the various environmental concerns. Of course, the relative importance varies not only from mineral to mineral but also from site to site. Our discussions with operators, Mineral Planning Authorities (MPAs) and other regulatory and interested organisations were helpful in complementing the literature. Together they give a reasonable and up to date picture of the relative importance of concerns and problems.

Various authors in the more general references describe concerns and complaints. For each report in turn and in broadly descending order of significance, these are:
- traffic and the inappropriate siting of plant[15],
- noise, dust, landscape changes, run-off, traffic[34],
- noise, blasting vibration, traffic, visual effects, ecology[431],
- traffic, blasting vibration and dust[433],
- noise, dust[439].

2. Concerns

A general survey[267] by WS Atkins based upon the number of local authorities reporting complaints found that dust, noise and traffic were equally bad in terms of nuisance[a], and that noise was the aspect most noticed by the public. It was one of the main causes of complaint in over 50% of cases around opencast coal and hardrock sites and in somewhat less than 50% of cases around sand and gravel sites. However, noise was rarely the main source of nuisance and complaints tended to arise mostly in the early operational stages. Dust created the greater annoyance and blasting was a problem where used. The reality of the effects of dust and blasting was often worse than people anticipated, whereas that of noise, traffic and visual effects was better. Visual effects were rarely complained of except at the planning stage. Less than 20% of those living near to surface mineral workings in Cumbria and Northumberland mentioned the workings as a negative feature of their area.

Devon CC's survey of their parish councils[134] showed that the main concerns were visual intrusion, traffic, dust, noise, blasting and footpaths although the number of complaints was low. A review of the mineral policies of National Parks[199] found that the main concerns were for visual effects and traffic and at a lower level for noise, dust, ecology/wildlife, heritage, socio-economic effects/residents, water and lastly rights of way and waste. Of the 25 Mineral Local Plans that have been prepared so far, the operational environmental topics mentioned in 10 or more were[485]:- transport, water pollution/drainage, noise, buffer zones, phasing of work.

An MPA with predominately sand and gravel workings recently surveyed residents' reactions to surface mineral workings. The 4% who considered themselves or their homes to be affected said that traffic and, to a lesser extent, noise were the main problems. Two thirds of all respondents expressed concern for water, landscape and ecology/habitats.

Our discussions, site visits and the written responses indicated a widespread concern at a professional level for groundwater and contamination of surface water by solid material and less frequently with oil and tar in discharges from workings. Other issues of concern are 'flashier' water flows following the removal of vegetation; dewatering leading to derogation of wells, reduced flows in streams, over-draining of farms and sometimes settlement; and odour which sometimes occurs from coating, other plant and in one case from the burning of waste oil to provide process heat.

2.1.2. Concerns about Specific Minerals

The concerns about **opencast coal**[14] mining are said to be traffic, noise (from earthmoving, reversing signals, rock drilling, construction of overburden mounds as screens), dust, blasting vibration, visual intrusion of overburden mounds, the effects (including lighting) of unsocial hours of working, proximity to housing, eg 50 m, and duration of workings, eg up to 22 years. Other sources report concerns relating to:- traffic, noise, dust, loss of rights of way and obtrusive impact[453], - blasting, dust, noise and traffic[35].

---

a. Unless otherwise indicated the word 'nuisance' will be used in the colloquial rather than the legal sense.

A sub-section of the Atkins survey[267] covering 32 opencast coal sites found that complaints reported by local authorities were of noise, dust, traffic and blasting.

A survey[324] of residents near to three opencast coal sites (proposed, current and complete) showed that moderate to strong opposition was expressed by 66%, 48% and 36% respectively of those interviewed. Opposition related to:- the view, loss of habitats, noise, blasting, fall in house prices, traffic and dust; opposition decreased as the workings progressed.

BCOE's records of complaints[71,75] over the last two years show an average of one complaint per month for each site. Complaints of blasting predominate, noise, other effects (including traffic), dust and water give rise to fewer complaints. It is suggested[230] that the number of complaints is increasing but that the total is still low.

The main problems with large **stone quarries** have been said[19,353] and observed to be visual, particularly scars on the landscape and unsightly waste heaps, traffic, dust on vegetation and dust affecting the clarity of the view in scenic areas, noise and vibration. Blasting in limestone quarries is said to give rise to widespread concern[454], concerns about noise, dust, traffic and landscape are usually less severe. A large limestone quarry in a National Park was said[353] to be visible and noisy over a wide area and to be the main source of heavy goods traffic in the area to the detriment of walkers and tourists. The sub-section of the Atkins survey[267] based on 41 quarries found that complaints were of dust, traffic, noise, blasting and visual effects.

Other adverse effects are reduced amenity of footpaths. Water pollution is generally not reported as a problem except where run-off is contaminated with fuel, dust, etc; also where quarries are near to caves, there can be paths for polluted surface water to reach the groundwater. Their longevity and the propensity for quarries to be extended are significant factors.

Specific data about **sand and gravel workings** is limited. The Atkins survey[267] based on 35 sites reported, again in descending order of significance, that traffic, noise, dust and visual effects were the main sources of complaint. The MPA survey referred to in Section 2.1.1 found that traffic and noise were the main sources of complaint.

Roadstone coating plants sometimes discharge unburnt fuel, tar, odour and smoke to the atmosphere and contaminate surface water with oil and tar. Some peat workings cause settlement of nearby roads and properties and affect the ecology of adjacent peatlands. Smell from fletton brick making[433,434] has been reported.

2. Concerns

2.1.3 Summary

Taking a 'global' view, the main concerns and problems are, in broadly descending order of importance:
- the various effects of road traffic, which create almost universal difficulties when this is the primary means of transport,
- at a similar level, but more selectively according to type of mineral and amount of overburden, the effects of blasting, noise and dust,
- visual/landscape effects, which produce more objections at the planning stage but fewer complaints during operation,
- contamination of surface water discharges by solids is relatively common, contamination by oil and its derivatives is less frequent; dewatering also creates difficulties with derogation of wells and streams, over-drainage and occasionally settlement.

Other issues, ie wastes, and effects on archaeology, agriculture and forestry, ecology, whilst they are potentially important issues, are not usually the subject of public complaint to Mineral Planning Authorities (MPAs) or to local authorities. There is less reference to such problems in the literature and it is unclear how frequently they occur. They are nonetheless of serious concern to MPAs, the National Rivers Authority (NRA), English Nature, Countryside Council for Wales and other such interested bodies.

2.2 MANAGEMENT OF CONCERNS AND CONTROL OF PROBLEMS

Before dealing with individual environmental concerns and problems, it will be useful to consider the different forms of planning condition and some of the types of action that operators might take as good neighbours.

2.2.1 Planning Conditions

If planning conditions are to be applied to permissions, they must be enforceable. This means that they must be:
- precise,
- capable of being monitored, ie infringements must be detectable,
- defined sufficiently for breaches to be provable.

In addition, to be valid[26,423] they must be necessary, relevant to planning and to the development, and reasonable.

They can set requirements in a variety of ways. The principal ones are:
- performance requirements,
- the use of specific ameliorative measures,
- the use of 'good' practice, eg as set out in a code of practice.

2.2.1.1 Performance Requirements

In the literature[371,485,486] and in our discussions with operators and planners, unease was occasionally expressed about subjective criteria being used. We can readily understand that those accustomed to engineering measures, eg levels of vibration likely to damage structures, may find it difficult to accept criteria designed to limit subjective reactions from residents. However with the exception of water quantity and quality, damage, safety and some physical effects of dust, most environmental concerns are subjective.

Concerns for noise and landscape/visual intrusion are virtually entirely subjective. Concerns for blasting vibration and overpressure, although exacerbated by a usually unjustified fear of damage, are mostly subjective. Concerns for traffic are not just the risk of accidents, delays and damage but also its environmental effects, eg noise, dust, visual intrusion. Any objective/numerical measure has to be based on a known relationship between a physical effect and the subjective reaction to it, eg the relationship between the level of noise and the amount of discomfort or complaint. The knowledge of such a relationship permits an informed but still subjective judgement to be made about the amount of complaint that will be tolerated. BS 4142[299] provides a relationship between noise level and the amount of complaints that are to be expected. Unfortunately, we found that some people are labouring under the misapprehension that such a 'standard' provides an objective limit. The problem is that most of the topics of concern have not yet been sufficiently researched to yield a robust relationship between levels of the effect and the subjective reaction of those exposed to it.

Part of the problem is that, whilst many operators are happy to see MPAs making subjective judgements on such issues, others are not and would rather have nationally agreed 'standards'. Whilst we suggest topics for further Departmental guidance in subsequent chapters, we believe that there cannot be effective planning without subjective judgements based upon the local situation and local objectives.

Where they are feasible, we consider that performance requirements are in many ways the ideal planning condition. They make clear to operators what is expected of them and leave them to decide the most cost-effective way of meeting those criteria. To decide methods for operators may inadvertently and unnecessarily prejudice the flexibility of their working methods and profitability. Performance requirements will usually be designed to achieve a minimum environmental quality or to limit degradation of the environment. An example of such criteria is a maximum acceptable level of noise at sensitive properties and/or other appropriate points.

2. Concerns

There are however sometimes difficulties and disadvantages. In our view, it must be made clear in setting performance requirements, that this does not sanction adverse effects for which there are readily available solutions and which do not involve significant costs. In some cases the operator will be able to achieve worthwhile improvements on the performance requirements without incurring significant expense. Clearly it is desirable that this be done but it is difficult to provide the appropriate incentives.

Some planners claim that performance requirements are unenforceable until they are exceeded[454]. It is maintained that continuous monitoring is necessary to pick up breaches of the criteria, eg levels of ground vibration from blasting, some have concluded that it is better to specify the maximum instantaneous charge or other ameliorative measures. However for most operators, periodic checks should be sufficient to identify undesirable trends and allow action to be taken to avoid breaching the requirements. In sensitive situations the provision by the operator of continuous monitoring systems can be a condition of the planning permission. Access to the monitoring positions is obviously essential and this will influence their choice. In may be desirable that monitoring be carried out at a sensitive property not owned by the operator; however, whilst criteria applicable to that property can be set in a planning condition, it cannot be a requirement that it be a monitoring position because the operator has no right of access. The choice of monitoring positions is discussed more extensively in Chapter 5.

The reluctance to be involved in monitoring and the difficulty or impossibility of defining a criterion which can be readily measured and enforced, eg for dust and visual matters, may lead to planning conditions which require the use of specific ameliorative measures.

2.2.1.2 Particular Ameliorative Measures

In order to reduce visual intrusion, the planting of trees may be required before mineral working begins or it may be specified that processing plant be moved into the quarry away from the entrance as soon as there is room to do so.

Under the Road Traffic Act 1986, the Construction and Use Regulations require that vehicles do not deposit material on the roads. The objective is simple enough to understand but it is difficult to enforce because the Police have virtually to witness or have evidence of a particular act against a particular driver/vehicle. In such a case it may be much easier to require specific action, eg the cleaning and sheeting of all lorries prior to leaving a site.

In these two cases of visual screening and sheeting of lorries, the condition is relatively easy to monitor, nonetheless there are still difficulties. Planting has to be effective in terms of screening and has to survive. Examples exist where planting has taken place but is ineffective. The cover sheets of vehicles have to be in good repair and effective. There is a risk that in trying to avoid these difficulties, planning conditions become overly detailed.

Other requirements, such as setting an upper limit on the size of blasting charges, may have cost implications for an operator and not be necessary, except when blasting is taking place close to properties. There should be an exception for any action taken in an emergency to avoid an accident. Such requirements might overlap with the responsibilities of the Health and Safety Executive (HSE) and may conflict with site safety requirements (see also Section 2.2.1). Where planning considerations have a bearing on the issue there seems to us to be no insuperable difficulty in setting conditions for matters which are the responsibility of other agencies. Indeed we see many advantages in drawing together requirements into a planning permission; naturally proposed conditions must be agreed in advance with the relevant agencies, eg HSE, the National Rivers Authority (NRA), etc. MPG2[26] does however advise (paragraph 61) that this approach should not normally be used but in paragraph 62 it also says that the imposition of planning conditions can ensure that any undesirable effect of development is mitigated or forestalled, whereas other legislation will often be corrective, with action possible only after the event. We wonder whether there is not room for a more integrated system for the assessment of planning applications and for the drawing together of conditions issued under the various pieces of relevant legislation.

2.2.1.3 Best Practicable Means

The use of 'Best Practicable Means' (BPM) is sometimes specified in conditions although 'Best Available Techniques Not Entailing Excessive Cost' (BATNEEC) is replacing this concept in some industrial sectors. These concepts are difficult to enforce in the minerals sector because it is unclear what they mean in any particular situation[267]. Her Majesty's Inspectorate of Pollution (HMIP) has defined what BPM involves for specific industrial processes and has periodically up-dated the definition in the light of technical progress[136,197]. This is not the case for mineral sites except for dust from processes prescribed for air pollution control by Local Authorities; for these the Secretary of State has issued guidance on what would constitute BATNEEC[491,492].

BPM and to some extent BATNEEC apply to pollution of a particular kind, eg air pollution. In many situations removing one form of pollution, eg solid waste disposal by incineration, may create another, eg air pollution. This gave rise to the concept of 'Best Practicable Environmental Option' (BPEO)[138] within which effects on land, water and air are considered simultaneously. To some extent BATNEEC is the EC's version of BPM[137] and to some extent it is BPEO[490] within the concept of Integrated Pollution Control (IPC).

There is little doubt that equipment such as hoods/silencers attached to machinery and provided for the control of noise ought to be properly maintained and used regardless of whether a maximum noise level is exceeded or not. Whilst the use of a different, quieter and more expensive machinery may be desirable it may not be agreed by the operator. Any uncertainty on BPM or BATNEEC would ultimately have to be resolved by the courts, but could be reduced by the use of accepted or designated codes of practice for mineral sites similar to BS 5228 'Noise control on construction and open sites'[284]. They would need to be reviewed periodically in the light of technical progress.

2. Concerns

Indeed one of the main advantages of this type of approach is that they can be updated without great difficulty. The Town and Country Planning (Minerals) Act 1981 allows the review of planning conditions (see also MPG4[53]) but any tightening of the conditions leads to the possibility of some compensation being payable to the operator.

2.2.1.4 A New Approach?

The suggestion has been made[393] that a two tier approach would be useful. In this there would be some monitoring, albeit relatively crude and simplistic. When a 'trigger level' is reached, which would be below the level of significant nuisance, a graduated response to avoid a potential problem would be set in train. The ameliorative methods to be employed would be agreed in advance and would be specific physical actions as described in Section 2.2.1.2. It might be useful if the trigger level and the graduated response could be reviewed in the light of experience; unless the concepts of BPM or BATNEEC were to be incorporated this would be difficult without raising the issue of compensation. Contingency measures could also be included for exceptional circumstances.

We feel that the approach could be useful where there is not enough confidence to make a numerical limit the basis of a planning condition, eg for dust, visual intrusion, traffic. The approach could also reduce some of the perceived difficulties of continuous monitoring (see Section 2.2.1.1).

2.2.1.5 Discussion of Planning Conditions

Some operators and MPAs prefer to apply BPM rather than to devote resources to what they see as extensive monitoring to enforce performance criteria. Others find the concept of BPM too vague without a list of actions that should be regarded as constituting BPM.

For small workings, very short-term workings or short-term activities on any site it may be felt that the cost of providing monitoring for the performance requirements is not justified. Nonetheless planning conditions specifying performance requirements may prove to be useful as a back-up for enforcement purposes in the event of unexpected changes in the operator's working practices.

Ideally for longer-term workings it would seem that there needs to be a combination of all three established forms of planning condition. This would involve research in the development of codes of practice, how they might be updated and the means of their enforcement.

We are attracted to the idea of a hybrid approach based upon a trigger level, particularly for dust and other factors where it is difficult to set and measure realistic acceptable limits. Whilst it is difficult to see how the approach might readily be incorporated in a planning condition and enforced, the idea merits further consideration and development.

In the Chapters which follow, the subject matter of planning conditions for the various environmental concerns is suggested. However, there is no attempt to draft legally acceptable wording.

2.2.2 Voluntary Measures

2.2.2.1 Compensatory Measures

There may be cases where the operator considers it economically unreasonable to carry out further amelioration at 'source' but where the effect on neighbours is still unacceptable. In such cases, if planning permission is to be obtained, the operator will need to consider other methods of reducing nuisance. Providing compensatory measures off-site, such as sound-proofing of houses, or compensation by house purchase, payment or some other compensatory action (see Section 12.4 on 'Compensatory Measures') can be useful ways of meeting performance criteria.

2.2.2.2 Non-Statutory Duties

There are a number of actions that we feel the various parties should undertake regardless of the existence or otherwise of a statutory requirement.

> As good **planners**, the Planning Authorities should consider what they could do to protect the surface workings from encroachment by sensitive land uses during its life, ie to keep them apart (see Section 12.7 'Buffer Zones and Local Plans').

> As good **neighbours**, the operators should use other better practices which do not entail unreasonable cost to minimise any adverse impact on the area, eg noise, dust, ecology etc. They should also keep open lines of good communication with their neighbours.

> As prudent **commercial organisations**, operators should monitor their performance and anticipate difficulties rather than wait for complaints or enforcement actions. It is arguable that there ought to be some sort of 'duty of care' in order to relieve the pressure on regulatory agencies. This topic will be discussed further in the section 'Monitoring and Enforcement' in Chapter 14.

To succeed, either as a good neighbour or a commercially prudent organisation in ameliorating environmental effects, it seems to us that an operator should ensure that the will to do so comes from the top down. This could be encouraged by seminars, sending operatives on courses, and initiating an environmental management plan.

2. Concerns

2.3 RELATED ISSUES

At and prior to the planning application stage, it is understandable that there is public resistance to change. For a new mineral working or a major extension to an existing site apprehension may well exaggerate the anticipated problems. Such apprehension can often be reduced, both before and during operations. It helps to ensure that those affected have contact with the operator and are given some understanding of what is involved in mineral operations and any compensatory measures that might be taken.

On the other hand, exposure to significant unresolved problems over a period, especially if it is longer than anticipated, may produce an increasing frustration. There is significant reference in the literature to a negative public attitude resulting from a lack of compliance with and enforcement of planning conditions designed to minimise environmental effects.

There may also be a cumulative effect. A number of effects due to a working, eg noise, dust, traffic, may individually be acceptable, however in total the situation may be unsatisfactory. A series of individually tolerable workings in an area may together produce an unacceptable situation.

It is also possible that the changes in the nature and quality of the countryside resulting from surface mineral workings and the type and quality of reclamation are significant factors in determining the level of complaint about operational aspects.

Planning gain, per se, is outside the scope of this study. Clearly, if it is seen as producing an advantage in the reasonably near future, it can beneficially influence public perception, as well as possibly offset some operational environmental problems. Planning gain deserves consideration during reviews of mineral planning policy and when debating the merits of an individual application.

These topics will be discussed further in Chapters 12 and 14.

2.4 RECOMMENDATIONS FOR FURTHER ADVICE AND RESEARCH

(1) It is not clear, in terms of environmental amelioration around workings with a life of a decade or more, how it can be ensured that advantage is taken of technical progress, without compensation on the part of the regulatory authority, unless BPM or BATNEEC is applied.

Further consideration, in relation to the longer term workings, should be given to ways of ensuring that advantage is taken of technical progress in ameliorating environmental effects.

Among the more obvious possibilities are:
- codes of practice,
    - their production and periodic revision, eg by individual operators or industry associations,
    - their voluntary adoption,
    - their designation under appropriate legislation or in planning conditions,
- a 'simple' procedure for determining what operational practice can be regarded as reasonable (eg, representing BPM or BATNEEC) in any particular case and at any particular point in time.

(2) In view of the variability in the environmental effects of working the various minerals, it should be a medium-term aim to develop codes of practice for each of the main minerals encompassing the relevant environmental concerns.

(3) The choice between the various types of planning condition is by no means clear cut. Ideally and for the foreseeable future, a combination of types is likely to be necessary. We suggest that research be carried out on the effectiveness of trigger levels coupled with contingent actions as a means of planning control.

CHAPTER 3

TRAFFIC

3.1 GENERAL

The potential off-site effects of traffic are to:
- add to the number and size of vehicles on the road; this may cause congestion, accidents, difficulties for pedestrians,
- damage roads or their verges,
- spill or drop material onto roads and spread dust,
- create visual intrusion, air pollution, dust, noise and vibration in areas adjacent to the roads.

On-site the potential effects are largely noise and dust in neighbouring areas. These issues are discussed in Chapters 5 & 6.

3.2 PROBLEMS

Many local authorities and others express concern and/or report complaints[14,34,35,267,427,433,453] about traffic regardless of the type of mineral. Even parts of the industry admit to traffic being the most intractable problem[339,464]. Complaints result from 'intimidation' by large vehicles, danger, use of roads unsuitable for the size of vehicle, damage to verges, dust, spillage, mud from wheels and body, noise from early starts and early arrival at sites, eg 06.00 hrs or Sunday evening and parking ready for Monday morning, vibration, and congestion.

Vehicles carrying minerals on roads local to mineral workings are among the heaviest and possibly the largest to use the roads in question. They are often out of scale with the rural and urban roads that they have to use, especially in the vicinity of the workings and the customer's site, eg the access to one major site is along a narrow road through a residential area with limited visibility and passing places. Even low flows in sensitive areas give rise to complaints.

Empty lorries are worse in some respects than fully loaded ones; they tend to travel faster and be noisier because they suffer from body-slap when going over bumps, road humps or sleeping policemen. If not sheeted, turbulence in the empty bodies of the vehicles may scour out dust.

3. Traffic

Some operators are under great commercial pressure to work 'unsocial' hours. Building contractors in urban areas require early delivery of aggregates or pre-mixed concrete to avoid delays during the peak hours and Highway Authorities need materials for the start of their working day, including weekends. This reflects the pressure to repair and maintain roads at the weekend outside the normal working week and especially to avoid congestion during peak hours. Indeed examples have been quoted of coated-roadstone plants starting work at 3 am in order to produce material to be collected at 5 am, even on Sunday mornings. These pressures are likely to be increased now that the Department of Transport is 'charging' contractors 'lane rental' for carriageway occupation when repairing roads.

3.3 ACCEPTABLE LEVELS AND MONITORING

There are no generally agreed levels of traffic along any specific road that are considered to be environmentally acceptable. Whilst some attempts have been to develop the concept of 'environmental capacity' for urban areas[479], the only readily available yardstick remains the purely technical one of road capacity. In practice, statistics of road damage, repair costs and accidents, etc are an indication of physical limitations, but we found no useful quantitative information about environmental limitations on road use.

A significant increase in the flow of heavy vehicles can be given as one of the main reasons for the refusal for a planning application[150,350]. In one case[350], although the additional flow was within the physical capacity of a B-road, it was described by the Inspector as representing a serious and unacceptable impact on the environment and amenities of the affected villages. The vehicles were described as large and intimidating and the road conditions as sub-standard in terms of alignment and visibility. The problems were exacerbated by on-street parking, narrow pavements and property directly fronting the highway.

There is a tendency to say that what happens on a public road is entirely a matter for the police[266] and that problems of spillage, etc are covered in the Road Traffic Regulation Act 1984 and the associated Motor Vehicle Construction and Use Regulations 1986. The implication is that there is no need for MPAs to concern themselves with such matters, eg to insist that operators undertake to sheet their vehicles.

We do not consider that this is an effective approach and this report will consider such issues further. The legislation is not enforceable until spillages occur and it is easier to monitor and enforce a requirement to sheet vehicles than a requirement that there be no spillage[266].

3.4 GOOD PRACTICE

In general there is little that can be clearly described as generally accepted 'good practice' regarding traffic associated with mineral workings, as the problem in many areas seems to remain unsolved.

3.4.1 Alternatives to Road Haulage

The first question is whether road traffic can be avoided in sensitive areas by using alternative modes of transport or by routeing it where its effects are more acceptable.

Hertfordshire CC[400] have presumptions against haul-routes using public roads between the excavation and processing plant and the use of significant lengths of local roads to gain access to the major road network. They encourage the use of conveyors from the excavation to the plant, a rail head or, to a lesser extent, a quay as means of distribution.

The British Coal Opencast Executive (BCOE) and the British Aggregates Construction Materials Industries (BACMI) have policies of using, wherever practicable, rail, canal or off-highway haul roads[339,340,456,474]. There is also the possibility of using a conveyor to a railhead[6], a processing plant or a distribution point on a suitable trunk road[20] away from the site. The tendency towards larger workings may improve the likelihood of rail heads being provided for new sites[484,488] and grants are available under Section 8 of the Railways Act 1974, eg 50% grant towards railhead[369]. However in some circumstances the use of rail transport may make it more difficult to blend coals and the operational costs are not necessarily lower than those of road transport. The question has been raised[15] whether mineral planning authorities in conjunction with British Rail should be able to provide more than encouragement to operators when contemplating rail access.

In discussing the possibility of alternatives to road haulage, it must not be forgotten that they in turn will have environmental difficulties of their own. In some respects they may be worse; the relative total impact must be considered. Caution must be exercised to ensure that problems are avoided at the receiving end[339,463] of any link. Except for coal delivered to power stations, nearly all minerals are transferred to their final delivery point by road.

Perhaps the biggest difficulty in achieving improvement in the haulage of sand and gravel is the sensitivity of the selling price of the mineral to transport costs. Haulage represents half the price after a journey of about 30 miles[339,460]. The situation is less adverse for the higher value minerals such as coal and also for rock-based aggregates where economies of scale make it economically feasible to transport the mineral for long distances, eg, granite comes from Scotland to England by sea.

The noise and dust problems arising from traffic on site can be ameliorated by the practices discussed in Chapters 5 & 6. However alternative methods of haulage and materials handling can have more far reaching benefits in the longer term.

Over the shorter distances, especially from excavations to processing plants and to distribution points it is suggested[368] that pipelines, aerial ropeways and conveyors can more easily be integrated into the environment than roads. A greater use of pipelines to transport sand and gravel may be possible; pipelines are already used where suction-dredging is employed as the means of extraction. They are also used for the extraction and transport of china clay. In other cases the use of slurry pipelines[81] and aerial ropeways[209,368] seem to

3. Traffic

have limited or no application. The main problems with pipelines are an adequate supply of water on the site, water separation and disposal at the terminus.

Conveyors are more likely to be viable[366]. The tube conveyor which wraps itself around its burden is environmentally attractive as it would avoid the problems of spillage and wind-blown dust[397]; the use of conveyors capable of ascending steep slopes[366,382] would reduce the need for long haul-roads and the use of dumpers. The use of mobile crushers[126,149] facilitates the greater use of conveyors.

3.4.2 Route Planning

If traffic is inevitable then there are advantages in selecting the most appropriate route, however ensuring that it is used may be difficult. Some planning authorities are fairly successful with:

- ensuring that the design of access roads and junctions within the site boundary, reinforced by signs, points vehicles in the 'right' direction[359,489],
- negotiating Section 106 (previously Section 52) or other legally binding agreements[305] relating to routeing.

Other local authorities seem to find the problem intractable. The authorities who claim to be relatively successful in controlling traffic problems, receive sufficiently detailed complaints from the public that they are sure of their ground when dealing with operators.

Operators point out quite reasonably[427] that, if the road is considered unsuitable for their HGVs, the highway authorities should effect a general size and weight restriction. Mineral operators should not be singled out. The problems are not limited to 'mineral' lorries; one suggestion is that all HGVs should be restricted to using the most suitable roads[135].

Planning permission can be refused on the grounds that the resulting traffic would exceed the capacity of the road and create an unacceptable environmental deterioration. In considering a planning application the Highway Authority will be consulted. If the additional traffic is considered likely to overload the road network, then the alternatives are to refuse planning permission or to seek a Section 106 agreement to fund road improvements, eg a roundabout, improved alignment. In the event of an obvious 'rat-run', restrictions might be placed upon the use of that road in advance of mining operations.

Generally speaking highway authorities will probably not impose restrictions on the use of roads until they are convinced a problem exists. An exception will be where bridges might be unsafe for heavy loads. The Police would be consulted prior to a restriction order and, whilst they would probably support a safety or traffic based constraint, they might not support a restriction based upon purely environmental considerations. As a result such a restriction, if imposed, is unlikely to be policed.

A Section 106 agreement on routeing, although relevant to the development and often used[305,] may prove difficult to enforce[350] because of the problems of demonstrating non-compliance. A Section 106 agreement was considered to be an inappropriate method of traffic control by the Secretary of State when giving his 'Brown Lees' decision[360]; it was said that any agreement should relate to the regulation of the development or use of the land in question and that it might be more appropriate to seek a Traffic Regulation Order under the Road Traffic Regulation Act 1984. Where significant problems exist Highway Authorities have wide powers[491] to restrict or prohibit certain classes of vehicle from using particular stretches of road at specific times[303]. Section 2(4) of the Act gives them the power to specify through routes for lorries or to restrict their use in certain areas, in order to preserve the amenities of an area. Alternatively control could rely on another form of legally binding agreement offered by the operator[384,400].

Several operators have contractual requirements on their own drivers and clients regarding routeing, cleanliness & sheeting[191]. In one case the requirements are overprinted on the delivery note. At least one operator has turned away a client's vehicles for not complying. BCOE have a policy[457] of agreeing routes where public roads are utilised. It is reported[463] on the other hand that, in spite of an agreement on routes with one operator, lorries invariably take the shortest route and that the operator claims to have little control. It is said that even a willing operator may find difficulty in enforcing such requirements[427].

Communication with drivers by 2-way radio-telephone can help to redirect traffic in the event of congestion or an accident and to investigate any complaint. It may also help to minimise queuing by scheduling. CB radio[209] cannot be used where there is blasting because it may interfere with electric blasting detonators.

Although an agreed code of practice on sheeting (see Section 3.4.3) is likely to be fairly even-handed in its effects on operators in an area, a similar approach to avoiding rat-runs may be less so. Nonetheless, a consensus approach applying to all operators has a lot to commend it and it would be worth agreeing roads to be avoided on an area-wide basis.

We believe that operators, especially the bigger ones, could impose routeing conditions on, not only their own drivers and vehicles, but also on those of their hauliers and of their clients. At the moment the pressure to do so is one of 'enlightened' self-interest; although it is also possible to envisage a more formal basis for such action as has occurred in the waste disposal industry. The Environmental Protection Act 1990 requires that waste producers have a 'duty of care' relating to the disposal of waste off-site by those that they contract to do so on their behalf[478].

3.4.3 Limitation of Effects

If road traffic is inevitable in a sensitive area then its effects may have to be limited. Planning authorities have had some success with:
- limiting hours of work,
- requiring the provision of vehicle washing facilities, sheeting of lorries.

3. Traffic

Voluntary limitations by the operator of traffic during school journey times can be helpful[352]. Vehicle speeds could also be 'voluntarily' limited on unsuitable roads. Hours of work for hauliers are difficult if not impossible to control. Without excessive effort, MPAs can only control the hours during which the site gates are open.

Highway Authorities can limit the size (width, height and length) of vehicles permitted to use a particular road to avoid danger to the public or other traffic[303]; length is often important on narrow winding roads. Highway Authorities can also limit the period of the day that certain classes of vehicles can use stretches of road, in order to preserve or enhance the amenities of an area[303,491]. Such limits apply to all road-users not just surface mineral operators.

Instances occur of operators providing suitable overnight parking space away from sensitive neighbours and allowing loading during the evening to facilitate an early getaway. This has been found to reduce the early morning rush and queuing and spread the traffic load. Another idea that we heard of, but have no details, is a mobile roadstone plant for use at the motorway or road repair site. This would reduce the necessity for the operator's permanent roadstone plant to operate at unsocial hours and alleviate the associated traffic problems.

BCOE require in their contracts[352,456] that vehicles are washed and are properly prepared before leaving their sites. Other operators require sheeting of loaded vehicles[338]. Somerset CC, Dorset CC and North Yorkshire CC have similar codes of practice agreed with most operators in their counties regarding sheeting[323], although there are exceptions to the rule, eg partly loaded articulated vehicles, vehicles engaged in surface dressing. Some operators sheet vehicles carrying sand but not gravel. 'Coal-nets' are said to be easier to fix than tarpaulins and avoid the channelling of wind underneath them, however whilst they are effective in avoiding spillage it is not clear to what extent they control dust.

Guidance by the Secretary of State[491,492] on what constitutes BATNEEC (see Section 2.2.1.3) under the Environmental Protection Act 1990, Part I in relation to the control of dust from quarry processes, including those concerning roadstone and coal, indicates that in most cases sheeting or total enclosure and the cleaning of loaded road vehicles will be expected; in some cases there are similar 'requirements' on internal haulage. In many cases planning conditions make this a requirement[359,362].

We believe that operators, especially the bigger ones, could impose conditions on not only their own drivers and vehicles, but also on those of their hauliers and clients. Conditions should relate to cleanliness, sheeting, good maintenance (of silencers, tailgates, etc), times of arrival, and possibly maximum speeds.

There are some anecdotal suggestions that a larger number of smaller vehicles would be more acceptable in some cases. We entirely accept that, when heavy vehicles are using appropriate roads, it is economic both for the operator and for the highway authority in terms of maintenance[395] for the largest capacity vehicles to be used. However where the roads are narrow and perhaps winding as well, it is the size of the vehicles not the weight which causes damage to verges, passing difficulties and hazards to other road users. In such cases, an operator could consider the use of smaller vehicles, especially if the alternative may be the refusal of an application.

Vehicle washing could be improved, especially for vehicles which are loaded at the working face. Some operators have suggested that it is difficult to know which methods of washing are necessary before experience has been gained of the pick-up of mud on the site in question. We believe that in difficult cases the better of the wheel and body cleaning and washing facilities should be used and in combination if necessary.

Types of cleaning facilities which are available include:
- simple metal grids that shake a vehicle as it passes,
- wheel spinners,
- single sprays, or groups of sprays (possibly controlled automatically) cleaning the haul-road approaching the exit and thus indirectly the wheels,
- sprays cleaning wheels only or wheels and chassis,
- hurricane blasting sprays, multiple high intensity whole body sprays.

Any one of these can be doubled up or used in combination with others. It would also help to have longer lengths of surfaced roads on-site to improve cleanliness. The efficacy of cleaning facilities varies considerably; sprays on roads need to be turned off in freezing conditions; the efficiency of some wheel washers is reduced by driving through them too fast. However, no data has been found which would allow a clear choice to be made at the planning stage for a particular site.

3.4.4 Future Good Practice

We see no insuperable reason in the longer term why all vehicles entering or leaving a site and carrying materials that might result in dust or spillage should not be required to be 'sheeted', have an efficient tailgate and be clean. At present there are some practical difficulties, eg:
- it can be difficult to sheet lorries without staging to provide easy and safe access to the top of the vehicle; drivers have been injured falling from their vehicles whilst sheeting, but at least one car driver has been badly injured by rock falling from a lorry[323],
- at the point of delivery it may be difficult to find space to remove the sheets and more difficult still to provide facilities for replacing them,
- the driver may have to put on boots and dress for dirty conditions before removing the sheets, which is time consuming and thus a significant disincentive.

In the short-term, sheeting bays at the point of dispatch could be provided more frequently, as by Tilcon at Swinden Quarry[323] and at the quarries of Pioneer Aggregates (UK)[399]. Tailored sheeting can be provided for the individual lorries rather than making do with standard tarpaulin sheets and ingenuity. There is also reference[301] to sheeting permanently or semi-permanently attached to a vehicle which still permits easy loading. In the medium term, sheeting itself can be further improved. Mechanically operated sheeting ('Easysheets') which can be manoeuvred from ground level is available, although some operators consider that it reduces the payload too much. Whilst one example of such sheeting weighed around 250 kg and represented 1.25% of the payload of 20 t and cost £2,000, this was off-set by reduced turn-round times, no sheeting accidents and no need for sheeting bays.

The more widespread use of such devices following requirements in the Environmental Protection Act 1990 will probably lead to the development of an even more economic vehicle sheeting system, which would make it simple and easy to clean, load, unload and for the vehicle to remain sheeted whenever in transit.

3.5 RECOMMENDATIONS FOR FURTHER ADVICE AND RESEARCH

(1) Easier-to-use sheeting systems could be encouraged by the publication of a guide to the commercially available options. This might encourage the further development of cheaper and more convenient semi-automatic forms.

(2) The development of lorries/tippers which are quieter when empty, especially avoiding body-slap, should be encouraged[460].

(3) The effectiveness of the various types of vehicle cleaning facilities should be examined in a variety of situations and the results published.

(4) Means of simplifying legal undertakings with regard to routeing[427] and sheeting should be explored. The possibility of there being a 'duty of care' on the part of the operator in relation to the vehicles of others as well as to its own should also be examined. Guidance on these matters would be extremely helpful.

(5) The development of techniques for assessing the environmental capacity of individual rural roads and linking the results with work on urban roads. This capacity should take account of local conditions, the proximity of housing, the use of the road by pedestrians, provision for passing, etc.

SUMMARY OF GOOD PRACTICE – TRAFFIC

Good Practice for Mineral Planning Authorities

- at an early consultation stage, encourage alternatives to road traffic especially between an excavation and process plant,
- consider the need to agree or specify planning conditions relating to the:
    - site entrance, eg which way vehicles can turn,
    - provision of signposting,
    - sheeting of lorries before leaving the site,
    - provision of sheeting bays,
    - provision of information and instructions to drivers,
    - provision of adequate wheel/vehicle washing facilities (it will be wise to check the facilities are indeed 'adequate' in the light of experience of the site),
- liaise with the Highway Authority to limit the size, weight, or axle loads of vehicles using particularly difficult roads,
- seek legally binding agreements from operators and their successors to use acceptable routes, except when prevented from doing so by <u>force majeure</u>, where traffic problems might still arise in spite of planning conditions.

Good Practice for Operators

- seek alternatives to road haulage from excavation to processing plant or depot, eg conveyors,
- seek alternatives to longer distance road haulage, eg rail,
- avoid sensitive areas and the use of large vehicles in narrow winding roads by agreeing routes,
- require that their drivers and others use agreed routes, use washing facilities and sheet their vehicles where appropriate,
- offer a legally binding agreement on matters which cannot be satisfactorily covered by planning conditions.

CHAPTER 4

BLASTING – VIBRATION, OVERPRESSURE & FLYROCK

4.1 GENERAL

Blasting gives rise to a number of effects:
- vibrations transmitted through the ground and pressure waves through the air ('overpressure') shake buildings and people and may cause damage and nuisance. The effects of the two factors are difficult for even an expert to distinguish without instrumentation. However, the pressure wave may arrive after the ground vibration by up to 2 seconds over a distance of 1 km. The perception of both factors is likely to be stronger inside a building than outside,
- audible noise, because it is part of the pressure wave, occurs at the same time as overpressure. It may be augmented by the rattling of windows, etc, caused by the overpressure,
- flyrock is the name given to fragments of rock propelled into the air by the explosions. This is clearly potentially dangerous to people and property both inside and outside the site. 'Flyrock' in this report always means that crossing the site boundary,
- dust, which is discussed in the Chapter 6,
- fumes, which may be noticeable in confined spaces.

The levels of overpressure and noise can be significantly affected by meteorological conditions, eg temperature inversions. Areas in which levels are enhanced will generally be down-wind[254] and the increase in level about 5 dB. In addition, in advance of warm fronts and behind cold fronts the overpressure waves can be channelled towards small focal zones. Although these generally occur adjacent to steep temperature gradients it is wind shear that is the more important causative factor. These small focal zones exist for about 10% of the time at distances of 3-4 km from the site of the blast and can be in any direction relative to the wind; the overpressure increases can be 15 dB[221,254].

The need for blasting varies significantly between the types of mineral being worked. For sand, gravel, clay and peat workings it is rarely needed. For coal workings it is often, but not always, necessary to loosen or 'heave' the overburden. Most of the energy stays in the ground because the overburden is only loosened.

For hard rock it is necessary not only to loosen the rock but to fragment and move it away from the face of the quarry, because of this more energy is lost to the atmosphere than with

4. Blasting

coal so the overpressure will probably be greater. The basis of good blasting design is to achieve the desired degree of fragmentation in the rock safely and economically[258,483].

As a result of these and other differences in the blasting of coal overburden and rock, less explosive per unit volume is used for coal than for hardrock and levels of ground vibration tend to be higher for coal for a given explosive charge and distance. Specifically, for a given explosive charge at a given distance, the ground vibration produced from opencast coal workings is greater at the upper 5% level than that from quarries[438]; the variability is also greater. The converse is true of peak overpressure, ie its magnitude is greater for quarries than opencast coal workings for a given charge and distance.

4.2 PROBLEMS

As the need for blasting varies with the mineral so does the incidence of problems.

A survey of MPAs[12] showed that 58% of those areas with blasting operations received 'significant' public complaints in spite of planning conditions relating to blasting in most cases. These conditions included: restrictions on time and frequency (94%), no Saturday or Sunday blasting (c50%), no blasting on Saturday after 13.00 hrs and any time on Sunday (c25%), warnings required before blasting (25%+) and control of blasting practices (c20%). 64% of MPAs specified vibration limits in the ground near houses.

One consultant found[25] that 38% of the problems he encountered were of alleged damage, 48% concern for damage and 1% flyrock. His experience probably represents the more serious end of the spectrum of problems.

There is one reference[353] to fumes occurring in caves but this will not be discussed further.

4.2.1 Problems due to Ground Vibration

It is clear that the main reason for complaints of vibration is not usually actual structural damage but fear of damage and/or nuisance; the threshold of human perception is well below the levels currently required in planning conditions. Some operators seem to argue that, in the absence of damage, there is no problem[286]. It is true that good public relations and explanations reduce anxiety and hence complaints; nonetheless, complaints continue to be received and reflect significant disturbance whatever the reason and so potential nuisance to residents is a major planning matter. It has been reported[289] that the quarrying industry acknowledges that damage criteria are no longer adequate and that nuisance criteria must be adopted.

There is usually some familiarisation with a disturbance over a period of time and people gain confidence that their houses are not being damaged. Others, who may feel that their early complaints have been ignored, may become increasingly vociferous.

4. Blasting

It has been suggested[353,461] that care needs to be taken to avoid damage to caves. Caves may be SSSIs, of tourist interest, or used by cavers for recreation. No examples of actual damage have been identified during the study.

4.2.2 Problems due to Overpressure and Audible Noise

Overpressure may vibrate buildings but damage is rare[225]. More frequently it adds to the perception of vibration and causes complaints by making windows, ornaments, etc, rattle[22] and startling people[336].

Overpressure is probably less of a problem[21,219,223] than ground vibration especially where the use of surface detonating cord and plaster/secondary blasting are avoided. Some doubt does however exist[253] about its significance (see Section 4.3.1 which discusses acceptable levels).

4.2.3 Problems due to Flyrock

Reported examples of flyrock occur about 20 times a year and half of these projections reach more than 200 m beyond the boundary[322]. Some maintain that this represents a 'rare' event[456] (in the case of opencast coal) but the Health and Safety Executive disagrees[459]. When it does occur it can be dangerous; whilst actual injury is fortunately limited to about 5% of instances, damage occurs in over a third[322]. Flyrock is thus a significant problem.

4.3 ACCEPTABLE LEVELS AND MONITORING

4.3.1 Acceptable Levels of Ground Vibration

The threshold of perception of vibration[14,270] is about 0.2–0.5 mm/s ppv[a] compared to a level of 12 mm/s for cosmetic damage[411], eg plaster cracks, and 50 mm/s for structural damage to residential properties[327]. The reference in MPG3[27] to 100 mm/s as causing damage was an error. The basis of these criteria[411] is not, as might be expected, the onset of cracking but an increase in the rate at which cracks occur; without blasting this was on average 12 new cracks per year in the test buildings.

Many MPAs are setting limits which are below the thresholds of structural and cosmetic damage in order to limit nuisance. For the 64% of MPAs specifying ppv limits[12], the majority of the limits were between 2.5 and 12.5 mm/s, 75% of these MPAs nonetheless received complaints. As 75% of the limits were set at 6 mm/s or less and as the average actual levels

---

a. *The abbreviation 'ppv' stands for the peak particle velocity of the vibration. It is not necessary for the present purpose to understand exactly what it is physically. The measure has however been found to correlate reasonably, if not perfectly, with damage to property and to the degree of human perception. There are many reviews of vibration and its effects*[224,270].

4. Blasting

were presumably below the limits, the critical level for complaints is likely to be less than 6 mm/s.

It is clear that compliance with these limits will mean that structural damage is very unlikely and, although cosmetic damage is possible, we will concentrate on the issue of disturbance. It has been reported that complaints are likely to be received when vibrations reach 1.5-2 mm/s[18,260] and 1.5-3 mm/s[206] and that people are more concerned about a few high level blasts than a greater number of small ones[205]. It has also been said that 3-4% of blasts lead to complaints whatever the level[327].

Cornwall CC[219] suggest on the basis of their experience that complaints will start to occur with levels from 1.6-3 mm/s during the day and 0.25-0.5 mm/s at night and that the majority of complainants are newcomers or the newly retired. Staffordshire CC report that 6 mm/s is found to be 'unpleasant'. Northumberland CC report[358] many incidents at 4 sites over 25 yrs which have lead to complaint and all have been at less than 5 mm/s.

The United States Bureau of Mines (USBM)[416] concluded that 5-10% of people would normally find 12-19 mm/s less than acceptable. They suggest that lower levels would be necessary to achieve a similar degree of acceptability where people are at home and subject to the rattling of windows etc, leading to fright and fear of damage and injury, eg at 12 mm/s up to 30% of residents might complain.

It has been proposed[18] that a planning limit of 5 mm/s be set and that 8.5 mm/s be regarded as the level at which significant nuisance begins. Others[1,2] have suggested limits as different as 12.5 mm/s and 2 mm/s. It has been reported[453] that oil companies are specifying 2 mm/s for vibroseis surveys. Attewell[233] has suggested, for non-structural reasons and specifically for blasting, 4 mm/s for houses and hospitals and 2 mm/s for old peoples' homes in rural areas. BCOE have variously proposed a maximum of 12 mm/s with 90% below 6 mm/s[474] and a maximum of 6 mm/s[477].

The interpretation of BS 6472[269] is said[260] to suggest that a range of 9-13 mm/s would be 'satisfactory'. This is inconsistent with the above experience about reactions to actual blasting. Some doubt the utility of the standard in the present context[455]. Clearly there is no consensus over desirable limits for disturbance due to vibration from blasting, although the trend in limits set in planning conditions still appears to be downward as more information comes to light.

The low levels which give rise to complaints are often contrasted with the levels which occur in dwellings as the result of the residents' activities[270]. People cause and accept higher levels of vibration by slamming doors and hammering than are caused by blasting, although the waveforms and durations are different[379], blasting vibrations last longer. Slamming a door produces vibration of about 17 mm/s[273] and footfalls about 0.5 mm/s. It is suggested[334] that footfalls below 1 mm/s are imperceptible and barely perceptible up to 5 mm/s.

The apparent inconsistency of people sometimes complaining of vibration levels below the threshold of perception may be due to the fact that most measurements are made in the ground and not in the floors of buildings. Resonance in buildings can amplify the vibration, this occurs particularly in timber floors and rafters. The actual response of the buildings, and hence people's perception, may be significantly different depending upon the type of building. There is also the possibility that people may be complaining of vibration induced by air overpressure as it certainly causes vibration in buildings[377]. Overpressure may even be the cause of some complaints attributed to ground vibration (see also sections 4.3.2 and 4.6.1).

Few studies have been made of the response of people to such vibration in real life situations. Most of the work on people's response to vibration, including BS 6472[269], has been based upon laboratory studies. In these studies there is no element of concern for one's home and peoples' fears are not always logical or based on fact but are nonetheless disturbing[358]. The effects of blasting have been described as "fearsome" especially to elderly people[358].

It is difficult to offer advice on suitable levels of ground vibration and overpressure because of the uncertainties in the understanding of the response to them. Levels will have to take account of local conditions and the nature of the workings. Ground vibration limits are likely to be in the range 5-10 mm/s and overpressure 120-130 dB if cosmetic damage and the worst levels of disturbance are to be avoided. Current permissions mostly require ground vibration limits in the range 4-12 mm/s although lower levels may be necessary for specific objectives.

It may be that permissible levels of vibration and peak overpressure from blasting should vary with the time of day in a similar way to noise limits. One suggestion that has been made[85] is that higher levels could be tolerated between 09.00-15.00 hrs, middling levels from 06.00-09.00 hrs and 15.00-20.00 hrs on weekdays and lower levels at any other time. The frequency of occurrence will affect acceptable levels.

We find the existing treatment of acceptable levels of vibration in BS 6472[269] unconvincing. Its draft revision[317] proposes some allowance for building structure type which may introduce difficulties in application. At least one operator would prefer to have a uniform standard[456]. Prediction of the response of structures is complex as one needs to consider the frequency spectrum as well as the magnitude of the vibration[408] in conjunction with the resonant frequencies and the rates at which vibration decays in the major structural elements of the building. A further complication is the varying sensitivity of people to vibration depending upon its direction. Clearly there is a need for research and guidance[455] on acceptable levels that can be monitored and enforced.

To avoid damaging caves and delicate stalactites in particular it has been suggested[353] that workings do not approach within 10-20 m.

4. Blasting

4.3.2 Acceptable Levels of Overpressure and Audible Noise

Damage in the form of broken windows[219] from peak overpressure is possible at 168 dB (linear with 'peak hold'[a]) or 158 dB[327] and unlikely at 140 dB[327]; rattling is possible at 120 dB[254] or 120-140 dB[219] (linear above 2 Hz). Limits for planning conditions have been set at 120 dB[18], 133 dB[222] and <133 dB[223].

USBM limits[253,415] are 134 dB for nuisance, at this level 5% of residents would be expected to complain that they were startled and frightened; even 120 dB will lead to rattling windows, feelings of annoyance and fright. The State of Ontario regard 129 dB as acceptable and 137 dB as unacceptable[253]. Other suggestions are that the vibration of wooden floors will be strongly perceptible over 130 dB[254] and that complaints are to be expected over this level[253]. It is suggested that, because of the variability in the levels from apparently similar blasts, a 95% confidence limit should be used rather than an absolute one[254].

It is reported that of 277 measurements of overpressure from various opencast coal sites none exceeded 124 dB[327]. As previously discussed, peak overpressures are lower for a given scaled charge in loosening opencast coal overburden than when fragmenting hard rock which leads to more venting[253,438].

It is suggested[219] that techniques for reducing vibration will also reduce overpressure and that modern techniques of blasting produce lower overpressures anyway. Others have also expressed uncertainty about the interaction between overpressure and ground vibration[455]. Doncaster MBC[287,403] however found that over a 5 year period the proportion of ground vibration readings over 2.5 mm/s reduced from 50% to 1.5% but that complaints had increased. It was concluded that this was due to overpressure because the complaints occurred in weather conditions likely to enhance overpressure.

A consultant, on the basis of very limited data, has suggested that there is little difference in effect between peak overpressure of 113 dB, ground vibration of 6 mm/s and a combination of 110 dB and 4 mm/s[223] – all he suggests are unlikely to give rise to complaints. He also suggests that it is not possible to set a level which will preclude complaints, so the best approach is not to set a limit and to rely upon the operator to do its best to minimise levels[223]. An operator[358] suggests that any limit would be unreasonable and a limit as low as 105 dB could not be complied with.

Limits have been set for unreasonable nuisance from the very short-term audible noise at 94 dB(A)[1,2], and 98 dB(A)[3]. To avoid potential problems, it is clearly preferable to blast at the noisier times of the day[225]. There is some suggestion that livestock may be affected by noise/vibration[1]. Noise is rarely a problem if the use of exposed detonating cord is avoided and as there is no need for this in blasting, the problem will not be discussed further.

---

a. Strictly speaking it is necessary to also specify the low frequency cut-off of the measuring equipment. Most research measurements are probably made with a cut-off of 0.1 Hz; in the field measurements are often made with a 2 Hz cut-off which will give readings about 1 dB lower.

4. Blasting

4.3.3 Acceptable Levels of Flyrock

No acceptable levels of flyrock have been suggested, other than to avoid it altogether outside the site.

It is said within the minerals industry that it is not possible to be sure of avoiding flyrock or of predicting when significant instances are likely to happen. Nonetheless, the Health and Safety Executive say[322] that 83% of flyrock incidents might have been avoided if the whole operation had been supervised by a person with a knowledge of explosives, blast design, controlled drilling and careful observation of quarry faces, and accepted blasting practices had been followed.

Use of the good practice described below should reduce both the severity and frequency of significant flyrock mishaps. They will not however be entirely avoided and buffer zones between the active face and sensitive areas will remain necessary. Guidance on 'safe' distances is needed from the Health and Safety Executive, possibly in conjunction with the Department of the Environment, along the lines of guidance given for hazardous industrial installations.

4.3.4 Monitoring

Equipment is available for monitoring both ground vibration and overpressure. Although relatively expensive, synchronisation of the instrument with the blast is not necessary and it can be left unattended; it can be set to only take readings when, for example, the ground vibration exceeds 0.5 mm/s or the overpressure 110 dB.

There is much discussion as to whether the limits should be based upon:
- a maximum level alone,
- a statistical level, below which a specified percentage of the blasts must fall,
- the two together, eg not greater than 12 mm/s and 95% below 6 mm/s[362].

Some operators dislike having an absolute limit because of the difficulty of prediction, especially for peak overpressure; some MPAs also dislike it because they report that there is pressure to monitor every blast. Unfortunately compliance with a statistical limit can only be determined by monitoring every blast and it is difficult to enforce; consequently some MPAs only use maximum levels.

Often, in the past, a statistical limit has been quoted without reference to the period over which it applies. Without such a period compliance or otherwise cannot be established until operations have ceased because the limit applies to all blasts during the life of the site. One suggestion[359] is that a suitable period would be 18 weeks; an alternative where the benches are worked across the site might be the period taken to progress one step. Any neighbouring property would then be subject to the worst and the lowest levels of vibration during the

Environmental Effects of Surface Mineral Workings

4. Blasting

monitoring period; it is possible that some blasts during the cycle would be sufficiently low not to trigger measurements.

Monitoring is usually conducted outside properties, although some operators and MPAs take measurements inside buildings, particularly following a complaint because ground vibrations can be amplified by the buildings by a factor of three. Measurement inside a building can also help to engender good relations with the occupants.

Records are helpful for quality control and reoptimisation of blasting design, consequently a number of operators monitor most, if not all, blasts. 'New' monitoring techniques are useful for evaluating and optimising blasting design[55,90,120,206], eg self-triggering sensors, multiple sensors, photography and computers along with various computer programmes, eg BLASPA[108], Sabrex(ICI), VIBREX[120]. Videos and portable digital seismographs have also been useful in analysing blasts[108,119] and establishing the pattern of vibration transmission for a particular site. This is still a difficult task and work is in progress to improve the basic technique of predicting vibration levels[408]. Computerised monitoring of blasts helps[218] to improve safety and efficiency, eg by reducing misfires, as well as controlling vibration.

4.4 GOOD PRACTICE

4.4.1 General

In general it is in an operator's financial interests not to blast where there is a viable alternative, eg mechanical ripping of overburden. Where there is a possibility of avoiding blasting, perhaps through new technology, this should be carefully considered in the light of environmental pressures. Historical precedent may not be a helpful guide to an appropriate decision.

Although ripping helps to avoid blasting overburden[373,379] it can be dusty and noisy. Modern hydraulic backactors have higher breakout capacities than earlier models and so can further reduce the need for blasting of coal overburden. They also provide better selectivity in excavation, ie they pick up less waste which has to be rejected in processing. Diamond wire cutting[376] can be used for slate working.

If blasting is necessary, the first level of good practice is to avoid unnecessary problems. One problem that can be reduced is the public's reaction to blasting. Concern for a person's home, particularly where they own it, could be reduced by a scheme of precautionary, compensatory and other measures which offer guaranteed remedies without undue argument or excuse[358] (see also Section 12.4). Having to fight for what are seen as rights only exacerbates the initial reaction. Examples are window cleaning and legally binding condition surveys; the latter can help to achieve 'hassle-free' compensation or repair. BCOE offers independent structural surveys to a proportion of householders within 200 m of its sites where blasting will occur, another suggestion[262] was that 25% of houses at this distance be subject to survey.

The inherent difficulty in using surveys is the interpretation of changes in crack patterns. Cracks open and close with the seasonal changes of temperature, humidity and drainage, and numbers increase as buildings age (see Section 4.3.1 above).

Nonetheless such measures would almost certainly reduce people's fears of blasting and indeed their adverse reactions to other aspects of the site's operations, even if they are never needed (see Chapter 12 for further discussion).

At a physical level particular care is necessary with a first blast at a quarry. It may otherwise give rise to abnormally high overpressure and vibration because there is no break face to give relief to the forces produced. Problems may occur at any time in the life of the working because of[403]: faults in the strata and other forms of heterogeneity; blasting in tight corners; blasting near to made ground; excessive burden; and manufacturers' tolerances/errors in explosives or detonators or operator's errors resulting in simultaneous detonation of more than one charge/hole/deck.

The means of controlling[21,90,223,225,271,274,281,289,290,396] ground vibration, overpressure and flyrock have many features in common and are used by the better operators[456]. It is said that many of the practices also aid cost-effective production[223]. Many of these measures are required for safety reasons by the Quarries (Explosives) Regulations 1988[152] and the approved Code of Practice[309]. Together these introduce a tighter regime which should reduce the incidence of flyrock and unusually high levels of ground vibration and overpressure[152].

The measures include the need for:
- correct blasting design including: a survey of the face profile[404] prior to design; ensuring appropriate burden to avoid over-confinement of charges which may increase vibration by a factor of two,
- accurate setting-out and drilling, including surveying as-drilled holes for deviation along their lengths and, if necessary, adjusting blasting design,
- correct charging, checking powder rise during loading – this is especially important in fragmented ground to avoid accidental overcharging,
- correct stemming will help control overpressure and flyrock; the control of ground vibration is probably also aided. Correct stemming includes:
    ○ not using drill returns but more appropriate purpose-chosen materials, such as 10 mm chippings mixed 3:1 with damp sand; the optimum particle size has been suggested as 1/25 of hole diameter[119],
    ○ controlling the length of the stemming column[119], too short and premature ejection occurs, too long results in excessive confinement and poor fragmentation,
- monitoring of blasting and re-optimising the blasting design in the light of results, changing conditions and experience.

There may be some means of control which conflict with each other and a compromise may be necessary to achieve a balanced control of the three effects.

4. Blasting

4.4.2 Good Practice for Ground Vibration

Ground vibration can be further reduced[90,225,274,289] by:
- allowing sufficient margins of 'safety' when designing from blasting trials of charge size versus ground vibration at given distances,
- using smaller maximum instantaneous charges (MICs); many individual charges fired at small time intervals only increase the vibrations by 50% of the first charge, the shock however lasts longer[1],
- using decked charges and in-hole delays which permit further reductions in MICs without increasing the number of holes to be drilled, which would be the case if bench heights were to be reduced instead,
- carefully designing the detonation sequence and inter-hole/inter-row delays[108,274]; delays should not be too small otherwise the explosions will effectively coalesce[410] and the benefits of small MICs lost,
- considering changing the frequency spectrum of the shock by varying blast designs in terms of the inter-hole delays[49,120],
- ensuring accurate detonation.

To avoid damage to caves and in the current absence of definitive information about the extent of the problem, it has been recommended[461] that buffer zones of 80 m be established around known caves and passages.

These measures may or may not add to cost. One report[85] is of tighter control having a "major impact [*increase*] on mining costs"; another of "considerable costs"[255]. Reducing bench heights below 10 m may cost significantly more[255]. Other reports[223,281] say that such measures actually improve blasting, fragmentation and safety and consequently the economic efficiency of the workings. One operator actually reported a reduction in explosive related costs, after introducing decked-charges to meet ground vibration limits. Overall it is difficult to disentangle any cost changes because of the considerable interaction between vibration control and effective fragmentation. There may be a clearer distinction where the requirement is merely to loosen the overburden.

In the general, non-mineral related literature, an oft quoted method of reducing vibration propagation through the ground is to use a trench as a barrier. Usually this is not practicable on a scale that is effective. One reference[379] does however suggest that, in the context of mineral workings it is an under-used approach, pointing out that the first box-cut is in effect a massive trench.

4.4.3 Good Practice for Overpressure

Overpressure can be controlled physically, but the debate continues as to whether it is practicable to use planning controls to monitor and enforce limits in view of the fluctuations caused by varying weather conditions.

Physical control[21,90,223,225,254,274,281,289,290,379,387,396,403,415] can be exercised by:
- not using surface detonating cord and, if it has to be used, covering it adequately,
- not using plaster blasting; achieving good fragmentation by blast design and if necessary by crushing,
- reducing the surface area subject to heave,
- reducing the degree of surface heave by minimising the total charge and using a low charge weight per delay,
- using an appropriate sequence of detonation and considering the orientation of the working face in relation to sensitive areas; if the direction of blast initiation is away from or at right angles to rather than towards a sensitive location, then reductions of 10-15 dB and 6 dB respectively may be possible[290],
- avoiding gas venting through local rock weaknesses (also a cause of flyrock) by accurate drilling and placement, and regular face surveys,
- ensuring that the trace velocity between holes is significantly less than the speed of sound, ie the delay between holes is more than 3 ms/m; this will avoid air-blasts from individual holes reinforcing each other,
- avoiding resonance with floors, etc of nearby buildings by using delays of less than 25-40 ms,
- not blasting in adverse weather conditions[289,290], especially significant temperature inversions[415], these include:
    - moderate to strong winds towards sensitive areas,
    - foggy, hazy or smoky conditions with little or no wind,
    - still cloudy days with a low cloud ceiling,
    - periods when the surface temperature is falling in the middle of the day,
    - periods when strong winds accompany the passage of a cold front,
    - before mid-morning or after sunset on clear calm days.

It is difficult to reschedule blasting because of the weather. It may need a whole day to load and fire the explosives and still to have time to deal with any misfire. One alternative is to have facilities for storing sufficient quantities of explosive on site in safe and secure facilities. Such a facility together with a significant buffer store of mineral prior to processing or delivery would however provide the flexibility necessary to avoid blasting in adverse weather conditions. Often explosives are delivered for each blast[403]. A cheaper alternative adopted by some operators is to rely on weather forecasts to schedule their ordering of explosives and thus avoid the need to store explosives on the site.

Clearly it would be advantageous to be able to predict the effects of the weather on overpressure in a quantified way[18]. The Meteorological Office has a computer programme which can predict the propagation of overpressure but sufficient local weather information to use it is unlikely to be available; to use the programme effectively needs the use of radio-sonde balloons[221].

4. Blasting

4.4.4 Good Practice for Flyrock

Flyrock is produced[271,322] when there is too much explosive energy for the amount of burden, stemming is inadequate, or the explosive energy is too rapidly vented through a zone of weakness. Geology, rock conditions, improper blasting design, or carelessness can cause flyrock. The operator should adapt blasting methods to suit the conditions when blasting in conditions that favour the production of flyrock, eg heavily faulted and fragmented rock.

The control[271,281] of flyrock is to a large extent accomplished by careful attention to detail. This may involve:

- avoiding misfires by all means available and in particular doubling up the initiation systems; taking great care when relieving misfires to ensure that an appropriate burden is provided[459],
- choosing the bench height, burden and stemming such that the blasted rock movement is primarily horizontal and outward and not upward,
- selecting the orientation of the face,
- inspecting the blast site for natural joints, bedding planes, mud seams, voids and any other weaknesses; inspect the working face for overhang/backbreak or raggedness before the blast hole pattern is laid out,
- ensuring that blast holes are not located close to points of weakness; this problem is reduced by lower bench heights,
- designing the blast to have efficient confinement, ie burden, stemming, delays and sequence of firing avoiding fuse lines being cut and late detonation by the detonation of adjacent holes; if delays are too long under-confinement of holes to be detonated later in a sequence may result, conversely if delays are too short over-confinement and blow-outs may occur,
- ensuring that, in multiple row shots, the delay between rows is long enough to allow the rock from an earlier row to move out so that the next row will have adequate relief. However the delay must not be so long that cut-offs occur and cause misfires that increase the burden on later firing holes,
- using toe-priming rather than collar priming/detonating,
- ensuring that drilling is accurate,
- checking the powder column rise during loading, not overloading due to cavities around the hole, not loading to near the collar of the hole,
- checking that all holes are loaded,
- avoiding secondary blasting,
- using screen nets to provide containment, in doubtful cases.

4.4.5 Discussion

The introduction of the 1988 Regulations has controlled the production of blast specifications and has acknowledged blasting as a branch of engineering which is more exact than has been thought possible in the past[458]. There is still much to do, although site trials of vibration from blasting are common[233], sufficient allowance is not always made for the uncertainties[408] involved in subsequent prediction.

It is maintained[222,225,254] that limits cannot be set for overpressure due to the variable results of blasting and changeable weather conditions. This is especially true over longer distances where the effect can be +/- 14-15 dB[221,254]. Similar arguments are also advanced in relation to ground vibration and flyrock.

Whilst the use of adequate margins of 'safety' in the blast design in order to avoid exceeding any set limits in spite of these uncertainties would seem to be an appropriate solution, it is argued that the cost of such safety margins would be excessive because the uncertainties are so large. It is suggested that a better approach would be to limit the number of occasions when ground vibration or overpressure exceeds a 'maximum' level. We cannot see that a planning condition which does not allow for day to day control by monitoring has any great merit. If one has to wait to the end of blasting or even a blasting cycle lasting 8 weeks before knowing whether more than 5 or 10% of blasts have exceeded the limit or not it may be too late to act. Planning authorities are not alone in wanting to know where they stand; operators have told us that they do too. We expect that the greater degree of control over blasting resulting from the 1988 Regulations should reduce the uncertainties and the variability of the environmental effects that safety margins have to cover.

In our view, planning conditions should relate where practicable to performance, ie they should set limiting levels for ground vibration and overpressure, rather than set out good blasting practice which is the remit of HSE. The operator may have more economic ways of achieving the same end or may be asked to do things which create an unsafe situation. The HSE should be consulted as a matter of course about flyrock. It strongly deprecates conditions which set limits on, for example, the charge per hole, because they may lead to unsafe practice and even the specification of a maximum instantaneous charge may cause an operator to work in an unfamiliar way and lead to an error.

In the case of flyrock the only valid performance criterion is to have no flyrock incidents. Although this should be everyone's aim, it clearly cannot be guaranteed. Some separation distance may be necessary to satisfy the Health and Safety Executive that there will be no undue risk to people and property. Buffer zones of 300 $m^{1,2}$ and 350 m^3 have been suggested with a ban on secondary blasting. For more consideration of local plans and buffer zones see Chapter 12. However there is a clear need for guidance on separation distances, especially with respect to flyrock.

4.4.6 Future Good Practice

Significant improvements in good practice are dependent upon having a better understanding of what causes nuisance to neighbours. Only then can further developments in the technologies of prediction, blast design and practice progress matters.

4. Blasting

4.5 RECOMMENDATIONS FOR FURTHER ADVICE AND RESEARCH

(1) Guidance on acceptable levels of ground vibration and overpressure is urgently required. It is strongly recommended that research be commissioned to explore nuisance caused by a combination of overpressure and ground vibration, bearing in mind that decreasing ground vibration by itself does not always reduce complaints. This work should to lead to advice on limits of ground vibration and overpressure necessary to keep disturbance to reasonable levels. It may be necessary to develop a new or modified form of measurement, possibly incorporating a combination of absolute limits and statistical variables, to improve the correlation of physical measurements with the subjective response.

This research will take a significant period of time. In the interim some outline guidance would be useful upon the factors that it is reasonable to take into account, eg type of property, nature of residents.

(2) It is strongly recommended that guidance be provided on separation distances relating to flyrock. The Department should discuss the matter with the Health and Safety Executive and encourage them to issue guidance or to set up procedures whereby planning authorities could obtain advice in particular cases.

(3) Consideration should be given to revising Circular 10/73 to include surface mineral workings and to include reference to vibration, overpressure and flyrock.

(4) The Department and the Health and Safety Executive should consider advising that the use of surface detonating cord be prohibited or discouraged and operators required to seek an exemption to use it.

(5) Consideration should be given by both the Department and operators to compensatory measures and procedures, eg, legally binding condition surveys to improve the possibility of 'hassle-free' compensation or repair, window cleaning.

SUMMARY OF GOOD PRACTICE – BLASTING

Good Practice for Mineral Planning Authorities

Consider the need to agree or specify planning conditions relating to the:
- levels of ground vibration and overpressure which can be readily monitored, preferably in the form of maximum levels,
- prohibition of the use of surface detonating cord and plaster blasting.
- control of flyrock, after advice from the Health and Safety Executive.

Good Practice for Operators

Many of the good practices, including staff training are required by the Quarries (Explosives) Regulations 1988. Particular care is needed near made ground, where there are faults in the strata or fragmentation, and in tight corners. Blast design should consider possible operative error with explosives and detonators and, to incorporate sufficient safety margins:
- carry out face surveys,
- design blast, including the size of MICs and detonating sequence, to minimise environmental effects,
- check the setting out of holes and record any deviations,
- revise the design, if necessary,
- use correct stemming,
- monitor the blast to provide feed-back for future blast designs.

To limit ground vibration:
- minimise MICs, eg by using decked charges,
- take especial care in unusual situations, eg in corners.

To limit overpressure:
- avoid use of surface detonating cord and secondary blasting where possible,
- minimise the area of heave and the total charge,
- avoid blasting in adverse weather conditions, especially when the wind is from the quarry towards sensitive premises and there is low cloud.

To avoid flyrock:
- ensure that the design is thorough and follows the Regulations,
- move fragmented rock horizontally rather than vertically,
- use toe rather than collar priming/detonation,
- use screen nets when in any doubt.

CHAPTER 5

NOISE

5.1 GENERAL

The potential effects of noise beyond the site boundary of a surface mineral workings are to:
- distract or annoy; a noise does not have to be loud to be intrusive, it may be different in character and identified as coming from an unwelcome source. Adaptation is possible, sometimes over a few weeks, when people accept the noise as unavoidable and believe that everything reasonable is being done to ameliorate it. If people are annoyed with the originators of the noise, it will be difficult for them to accept and to adapt to it,
- mask desirable 'noises', eg conversation; some adaptation is possible but this is likely to be at some cost, eg lost opportunities to sit in the garden, or the effort of close concentration,
- prevent or disturb sleep; adaptation is possible and likely in most cases, provided that there is no underlying irritation with the cause or with the originators,
- disturb animals and birds which can also be affected, particularly by sudden noises.

Noise has been the subject of a major study by WS Atkins[267]. This will be referred to extensively in this Chapter. The study found that the prediction of noise levels using the methods of BS 5228[284] were reasonably accurate, especially when using the specific measured noise values of the proposed or actual items of plant and equipment[267]. Predictions for the larger surface workings are often made on a computer[11,398].

Climate affects the propagation of noise. Calm weather often means a low background noise level and a uniform propagation in all directions. A light wind enhances levels downwind by about 3 dB(A); up-wind there can be significant reductions. With high winds noise propagation is variable; the background levels are likely to be higher and may mask other sources of noise.

5.2 PROBLEMS

Overall the actual problems of noise are usually less than residents expect them to be[267]. It is one of the main problems at about 50% of sites, but is rarely the main problem[267]. Complaints can be vociferous in the case of opencast workings but usually fade away[409].

5. Noise

Quarries tend to be better except when top-soil is being stripped. Unless reassured neighbours fear that this will be a regular problem and react accordingly. Extensions to existing quarries tend to receive fewer objections[128].

Early morning noise from vehicles arriving and on site plant being started-up is increasingly the subject of complaint[324,379,453]. Empty lorries can be noisier than loaded ones because of the 'body-slap' when going over bumps, including sleeping policemen. On cold mornings, night watchmen have been known to warm-up plant before the working day starts[378].

07.00 hrs has been a normal starting time for many years, however there is an increasing tendency for complaints at that time of day due to noise well below conventional daytime noise limits. There is also a tendency, especially at aggregate and roadstone sites for vehicles to arrive progressively earlier because of market pressures from customers (see Chapter 3).

Other causes of complaint are:-
- reversing warning signals[62,205,267,441,460] which are referred to as "the most significant additional noise intrusion over the last few years"[460],
- the squealing of dry caterpillar tracks and the operation of draglines[358], dredgers, dumpers, drills and pumps at night,
- the construction of screening bunds,
- earthmoving and the stocking of overburden mounds[378],
- fixed plant; this is a less frequent problem at opencast coal sites.

Noise from blasting is discussed in Chapter 4.

The construction of baffle banks or bunds and soil stores is one of the noisiest operations involved in mineral working[205]. As noise screens[84] are usually close to noise-sensitive dwellings their construction creates considerable nuisance. Bunds can cause difficulties in other respects. They can be visually intrusive[34], create dust, affect TV reception and may sterilise land and delay restoration[318].

Birds with poor breeding success levels are easily disturbed[449] whilst nesting and noise is probably a significant factor.

5.3 ACCEPTABLE LEVELS AND MONITORING

5.3.1 Acceptable Noise Levels

Peak noise levels are measured in dB(A) but should be qualified by the time-weighting of the meter, eg 'slow', 'fast,' or by the statistical level, eg exceeded for 1% of the time. Over a period it is common to average the noise in terms of its energy to give an equivalent

5. Noise

level (Leq), also in dB(A). The period over which it is averaged is the subject of differing views but should always be stated, eg 1 hour, 12 hours.

Acceptable intrusive noise levels will depend upon the ambient levels, the times of day, the duration (weeks, months, years) and the nature of the noise.

WS Atkins[267] suggest that noise levels at the worst affected house (or school, hospital, open area used for relaxation by the public) should not exceed:
- 55-60 dB(A) by day (typically 07.00-19.00 hrs),
- 45-50 dB(A) in the evening or dawn (where considered appropriate),
- 40-45 dB(A) at night (typically 22.00-06.00 hrs).

All levels are in terms of 1 hr L_{eq}; weekends and holidays can have different timings.

WS Atkins suggest that for permanent plant and haul-roads the daytime figures ought perhaps to be reduced to 50-55 dB(A). Conversely, the levels could be relaxed for short-term activities, eg the construction of overburden mounds.

It has been argued that the use of short averaging periods of 1 hour represents an unreasonable constraint on an operator[352] and that the use of longer averaging periods allows more flexibility and a trade-off between the quieter and noisier periods of operation[456]. In this case, we would suggest that a maximum level, as well as an Leq, be set to avoid unacceptably noisy periods.

Long averaging periods give rise to problems in monitoring and have lead to the suggestion of MPAs imposing conditions requiring the use of 'best practicable means' (BPM)[378]. This approach has the merit of allowing the benefits of improving technology to be reflected in reducing noise emissions from the longer-term sites. The point has also been made[379] that in setting noise limits, planning permissions do not sanction noise problems that can be readily solved at a reasonable cost. Enforcing the use of BPM is difficult because of a lack of definition; enforcing conditions which require the use of specific practices or which limit noise levels may be easier.

BCOE have suggested, following consultation with the MPA, that a daytime level of 65 dB(A) for $L_{eq,12\ hrs}$ near to sensitive properties would be acceptable[477]; in other cases they have suggested the same level as the limit at the site boundary[172]. They also suggest that stripping or replacement of top- or sub-soil, the construction of baffles or the outer faces of mounds should be exempt from such a restriction.

Somerset CC suggest[455] that, because of the very low ambient levels in some areas of the county, the night-time levels put forward by Atkins could cause problems. Somerset CC have used an $L_{eq,1\ hr}$ of 40 dB(A) for the night and expect complaints when the level is above 45 dB(A) but not below 35 dB(A)[261]. Indeed night-time working should be approached with extreme caution and avoided where possible.

Environmental Effects of Surface Mineral Workings

5. Noise

Any noise at night, particularly an intermittent one, may be the cause of complaint even if it does not wake people. This mitigates against the use of Leq over the whole night time period. It is even suggested[205] that the use of $L_{eq,1\,hr}$ in rural situations can be misleading as an indicator of nuisance, which is more likely to be determined by isolated noise events; more research may be needed.

The ambient noise climate and other factors vary significantly from one site to another and therefore we would emphasise that the levels and timings suggested by WS Atkins should be moderated to create site-specific standards. These should take into account:
- the pre-existing noise climate,
- the likely duration of the noise (weeks, months, years),
- the site operations in terms of operating hours.

The need to consider the existing noise climate, is demonstrated by BS 4142[299] which, in dealing with industrial noise in mixed industrial and residential areas, concludes that the level of complaint depends upon the difference between the industrial noise and the residential background level. It is argued that BS 4142 is not relevant to surface mineral workings which are more akin to construction work; this argument has had more effect on the approach to opencast coal than other mineral workings[128]. Nonetheless the need to relate acceptable levels of intrusive noise to the pre-existing background level is virtually unassailable even though the quantitative relationship between excess noise and complaints is not fully understood[409]. It is suggested[409] that a criterion of +5 dB(A) above the L_{90} level would avoid complaints. However according to BS 4142, a criterion of -10 dB(A) based on L_{eq}[299] would be needed to have a positive indication that complaints are unlikely; conversely at +10 dB(A) complaints are likely. These criteria could be relaxed for non-permanent activities.

Cheshire CC are reported[267] as having a policy which is based upon the existing background level and upon the capacity of an area to accept increases in noise level. In a purely residential area, by requiring the level from workings to be at least 10 dB(A) below the existing background, they effectively permit no increase in the overall level. In areas of mixed residential and industrial development or rural areas with scattered houses, Cheshire suggest that a small increase, 5 dB(A), in the background level can be accepted. Wherever the incoming noise is tonal, impulsive or irregular the acceptable level is reduced by 5 dB(A).

Hertfordshire CC require a $L_{eq,5\,min}$ by night of not more than $L_{90}-5$ and an $L_{eq,1\,hr}$ of $L_{90}+5$ by day[400]. These levels are adjusted for tonality and impulsiveness. This means that workings are effectively inaudible by night and that there is a small increase in the overall noise level by day. Pumps, etc, which operate 24 hrs a day are required to be inaudible.

Circular 10/73[297] seeks to avoid increases in the ambient noise levels at residential and other sensitive development. Both this document and BS 4142 are concerned with permanent sources of noise and the levels they suggest may not be appropriate for some of the shorter-term mineral workings. Nonetheless they indicate the levels likely to be appropriate for the longer-term workings and their associated processing plant.

5. Noise

The need for guidance has been expressed[455] as to whether to treat long-term surface mineral workings as 'permanent' industry or as 'temporary' construction works for noise purposes. It has been suggested[378] that the inclusion of opencast coal sites in BS 5228[284] might have been a mistake and should be taken out, presumably because inclusion implies they should be treated as short-term works and the noise requirements should be less onerous than for 'permanent' industrial development. The contrary view is also held[133].

Opencast coal workings are said[379] to be more continuous and less labour intensive than most construction sites. One report[409] talks of the possibility of a "spectacular concentration of heavy plant" for 10 years at opencast sites and of plant sitting in one place for 20-40 years at deep hard rock sites.

A level in excess of an L_{eq} 12 hrs of 70 dB(A) is rarely permitted around construction sites in an urban area. Under the Control of Pollution Act 1974 the local Environmental Health Officer has the power to set limits for construction works. It has been suggested[409] that, for surface mineral workings where BPM has been applied, 70 dB(A) be permitted but only for limited periods, eg 8 weeks in a year or 6 weeks in 6 months.

Whilst the above levels suggested by WS Atkins can be relaxed for limited periods for very short-term activities, eg the construction of bunds, we feel that they may need to be more stringent for a surface mineral working or a succession of workings which are effectively permanent from the point of view of nuisance (see Section 12.5).

General experience, eg with construction sites and opencast sites[379], suggests that, where contractors utilise their full day time noise ration early in the day, complaints are most prevalent (see also section on hours of work in Chapter 12). A later start to the day, at say 07.30 hrs[379], or a transitional period (as is often provided in the evening) between a low night-time level and a high daytime one, eg from 07.00-08.00 hrs, would avoid some of the early morning problems. One survey[324] indicated that residents near to opencast coal sites would regard work starting at 08.00 hrs instead of 07.00 hrs as a significant improvement. West Glamorgan[453] suggest transitional periods of 06.00-08.00 hrs and 20.00-23.00 hrs. As above, WS Atkins[267] have suggested a transition from 06.00-07.00 hrs. Another suggestion is for a later start on Saturdays[378].

In at least one case[352,360] work on baffle embankments, overburden mounds and soil dumping close to (100 m) property has been limited to weekdays 08.30-17.30 hours. In another, an operator starts 30 mins later when working near houses and 1 hr later when working on bunds.

In summary, we suggest that a Mineral Planning Authority will have to consider whether the noise from the site may cause a significant deterioration in the existing environment. When they believe that this is a possibility, they should set different limiting levels at sensitive locations for various times of the day and days of the week. These levels should take account of the existing environment, any planning policies or intentions they may have for the area and the likely duration of the noise. Levels could be specified as valid for an agreed

5. Noise

period (years) and any overrun or extension of this period could trigger a review of levels and a possible reduction in the limits.

We suggest that, for really short-term works, such as the construction of noise/overburden mounds, more relaxed noise limits be set for a limited duration (weeks). This would be better than exempting the activity from the normal controls. The situation is akin to those for which BS 5228[284] was originally conceived in dealing with construction works, eg piling.

5.3.2 Monitoring

So that there should be no misunderstanding about the requirements, we and others[409] suggest that it is desirable in most cases to specify acceptable noise levels at sensitive locations. Such levels should reflect what is desirable more than what is easily achievable[409]; to do otherwise will not lead to a uniform standard at sensitive properties.

In one case[396] the operator suggested that it would be more realistic and ease access for all concerned if the same points near to sensitive properties were used for the monitoring of noise, ground vibration and overpressure.

The acceptable levels will often be close to existing ambient noise levels and may be difficult to measure. It may be helpful therefore to agree proxy monitoring points nearer to the site where the operational levels will be higher and easier to measure; it will also help if the monitoring points are free from extraneous noise, eg of children playing. These proxy points can also be chosen to be more easily accessible and will often be on the site boundary.

There can be additional limits, eg Hertfordshire CC specify levels at the site boundary in addition to those at sensitive properties in order to protect uninhabited areas. These limits may be more or less severe depending upon the distances involved. Care will have to be exercised in calculating the appropriate corrections to take into account, not only distance, but also screening, ground absorbtion and possibly climatic conditions. It has been pointed out[455] that for longer distances, the weather introduces uncertainties which may result in unduly conservative or lax corrections. As the workings develop new proxy points may have to be selected. In most cases the corrections to the measured levels must be reviewed periodically to ensure that they reflect the changing geometry of the site and disposition of mobile and fixed plant.

A suggestion[267] that these proxy points and corresponding levels are included in planning condition seems to us to be inadvisable. Assumptions would have to be made about the method of working, the location of screening mounds and to some extent the type of plant to be used. With the best will in the world an operator's information, valid at the time of planning permission, may change. We recommend that planning conditions should specify clearly what is wanted in order to protect sensitive land uses. Proxy monitoring points can be subject to later agreement.

5. Noise

The means of monitoring the noise climate should be agreed between the MPA, the local environmental health officer if appropriate, and the operator. It may be advisable to use a common set of measuring equipment especially if it is to be 'permanent' or 'semi-permanent'. In the case of longer term workings it might be appropriate to require the operator to provide a 'semi-permanent' or even a 'permanent' monitoring system. Vandalism can however be a problem with permanent systems.

'Semi-permanent' monitoring equipment is fairly readily available but is sophisticated and expensive. For short-term workings it may not be justified; however, for many other workings it should be considered. The sort of semi-permanent equipment likely to be at the disposal of local authorities will be capable of being left unattended for up to 12 hours. Hand-held equipment is commonly available.

Whilst Atkins[267] suggest that levels be specified over periods of 1 hour, they also propose that 15 minute readings at different times of the day may be better than a limited number of 1 hour readings for general compliance monitoring. The equipment should be manned so that extraneous noises near the microphone, eg dogs barking, can be eliminated. Clearly if enforcement proceedings become necessary, 1 hour readings and possibly monitoring around the clock will be needed. The difficulty of eliminating extraneous noises, eg by having to man the monitoring equipment, mitigates against the use of noise limits based on long measurement periods.

5.4 GOOD PRACTICE

A great deal of literature on the avoidance and reduction of noise is available[84,101,226,260,267,282,283,284,384]. Good practice can be considered in terms of planning to avoid noise, controlling it at source and reducing its propagation.

5.4.1 Planning to Avoid Noise

Much can be done to minimise noise by good planning of the site in relation to neighbours and in terms of the way in which the site is to be worked (see Section 12.7 'Buffer Zones and Local Plans).

During the planning of operations, some improvement is possible without changing mining practice. A CIRIA report[282] aimed at reducing occupational noise levels, gives relevant advice on how to plan to avoid noise problems and extends the advice in BS 5228[284].

Primarily, noise can be avoided or reduced by thinking ahead and including noise as a factor when making basic decisions about design and operation. The basic elements are that:
- the client/operator accepts the responsibility for and cost of avoiding noise nuisance,

Environmental Effects of Surface Mineral Workings

5. Noise

- where tendering is involved, competition is fair; it should be a condition of tendering that the contract will not be awarded to a contractor/supplier who does not demonstrate that the bid has taken noise into account and responsibility for meeting noise limits is accepted,
- noise is one of the factors which determine the design of the workings and the sequencing of activities; Computer Aided Design can be a major help in doing this,
- the proposed method of working will meet any planning conditions related to neighbourhood noise,
- where a choice of methods or plant is available at reasonably comparable costs, the quieter should be chosen, eg for earthmoving plant this might produce a reduction of 5-10 dB(A)[385],
- no plant is considered before its level of noise emission is known (data needs to apply to comparable working conditions to that expected on site),
- the site is laid out to minimise the noise impact on neighbours, especially from significant noise sources such as the construction of overburden mounds and mobile or stationary plant in 24 hour use. It has been suggested that excavation should not take place nearer than 100 m to a dwelling[384]; such a limitation can only be site-specific.

Possible examples of such planning include:
- sequencing activities to utilise the screening of the working face[267] or overburden mounds,
- low levels of activity in the evening and no work at night,
- the siting of haul-roads so that they are screened by the topography, minimising road gradients to avoid low-gear high-revving driving, providing sleeping policemen to control speeds but taking care to avoid creating local sources of noise from vehicle body-slap or revving engines as drivers accelerate after passing over them,
- less haulage by truck and greater use of aerial ropeways[368] or conveyors[366,382] which are capable of climbing slopes of 60°,
- use of electrically-powered plant with its power source in an acoustic enclosure[364],
- use of low-profile plant to reduce overall height, eg in one case from 14.5 m to 9 m, will improve the screening by local topography and artificial bunds,
- processing plant insulated by acoustically efficient cladding.

5.4.2 Control of Noise at Source

Planning for noise avoidance or limitation should be coupled with effective site supervision including the proper use of equipment and good maintenance[282]. It is vital in this context that management objectives include operating the site as quietly as possible.

We suggest that plant noise levels should be established before it is allowed on site. No significantly noisy item of plant should be allowed to operate before its noise characteristics have been confirmed by measurement. The noise characteristics should also be checked at intervals in case there is a deterioration in acoustic performance.

The main sources of noise are[226]:- trucks, scrapers, loaders, dozers, whistles, sirens, screening and crushing plant. By night, pumps[379], draglines[379] (squeaking clutch bands and rattling chains), dumpers, drills and dredgers must also be included. One author[162] emphasised blasting and crushing. Reversing alarms are particularly irritating.

There are alternatives to the conventional 'bleeping' reversing alarms. One acoustic alarm 'listens' to the ambient level of noise and adjusts the level of the signal accordingly[331]; another[332] reduces its level when the side lights are on or switches off when flashing lights are in use. Somerset CC have reported that the Health and Safety Executive have agreed to the use of flashing lights at night in lieu of audible signals. Work is in hand to develop the use of close circuit TV to provide drivers with better vision when reversing[346]. One radar-based device provides the driver with audible and visual warnings which increase in intensity the nearer he comes to an obstruction; it can also be linked to the vehicle's brakes. However one series of tests[425] showed that the radar could miss obstructions if the vehicles was following a curved path and, when the vehicle was moving quickly, the braking was not always effective.

Noise reducing measures which can be applied to existing plant and facilities[282,284] are to:
- minimise the height which material drops from lorries or other plant,
- empty dragline buckets as near as possible to spoil heaps,
- minimise the clanging of dragline chains and buckets by careful operation,
- use rubber linings in chutes, dumpers, trucks, transfer points, etc, to reduce the noise of falling rock upon metal surfaces[109,260,369,383,439],
- use simple baffles around washing-drums[439], rubber mats around screening plant[347], crushing and coating-plant[370],
- enclose pumps[379], clad plant[369] and ensure that the cladding is kept free of holes, enclose conveyors; in sensitive cases ensure that the connections between the cladding and the plant is flexible to avoid undesirable transmission of vibration and noise,
- switch off equipment when not in use,
- start plant one by one,
- limit the use of particular types of plant and/or limit the number of items of plant in use at any one time[474],
- avoid unnecessary revving of engines,
- keep noise control hoods closed when machines are in use,
- maintain equipment to ensure the integrity of silencers, lubricate bearings, keep cutting edges sharp,
- point the noise away from sensitive areas whenever possible, if the noise from plant or equipment is highly directional,
- arrange to keep internal haul roads as smooth as possible,
- keep lorry tailgates closed where possible.

5. Noise

Where appropriate retrofitting of mobile plant can be arranged. Usually the first approach should be to the supplier or his agent as often they will have had experience of retrofitting their own equipment, and may have ready-made kits. This is important, as achieving significant noise reductions, will usually involve more than fitting a better exhaust silencer. It may include reducing the noise from the fans and enclosing the gear box and transmission. Reductions of up to 15 dB(A) have been suggested in general[385] and 8-10 dB(A) in a specific case[384] where the exhaust silencer was not already a good one; with good silencers the benefits of retrofitting would be slightly less.

The modification of plant is usually more likely to be successful when done by the manufacturer or his agent before the plant comes onto site[282] than if a DIY approach is adopted on site.

Stainless steel acoustically-absorbent materials are available[205] which are resistant to dirt and can be easily washed. Water has been used effectively to lubricate screeching chain links in wet and dirty conditions[205].

Some operatives have the mistaken idea that leaving hoods open helps to cool engines; in fact where they have been designed properly, this ruins the flow of cooling air and the engines are more likely to overheat. Other operatives consider that noise is synonymous with power or that they need a loud noise in order to judge the performance of their machine. All of this should be considered by site management as 'bad practice'.

5.4.3 Control of Noise Propagation

The main means of reducing noise propagation is to distance the receiver from the source and provide 'semi-permanent' screening by bunds, etc. When such measures are used they should be an integral part of a planned and integrated approach. Noise banks, bunds, barriers, fences and movable screens all have a part to play in reducing noise[284], but it needs to be clear why they are used. Not only noise but visual effects, TV reception (which can be tested in advance), land requirements, restoration and mining considerations all have to be taken into account.

Where bunds would be visually intrusive (residents have been known to decline an offer to provide bunds on the basis of visual intrusion[379]) it may be better to use noise barriers or fences[205,267] which can be coupled with planting. Trees alone, except over large areas, do little to reduce noise; they have a largely psychological effect. Where overburden has to be stored near to houses, a noise fence or barrier could be erected first and the overburden mound shaped to avoid further visual intrusion.

It has been questioned[205] whether mounds are preferable to fences or barriers as noise screens. They are much noisier to construct and less effective for a given height than barriers, such as wooden fences. A fence can be nearer to the receiver (or the source) and can be lower in height than a mound for a given noise reduction. The appearance of a fence can be improved by appropriate planting.

Noise effects from mound construction can be reduced if work on the mound face nearer to neighbours takes place during the noisier times of the day and work on the face farthest from neighbours takes place at other quieter times; the first section of a bund can screen work on later sections[379]. Where it is necessary to construct mounds near sensitive areas, the work can be screened by noise fences or portable screens[283] which could be made of straw bales.

The starting-up of major noise producing mobile plant should take place where possible behind overburden mounds or screens.

We consider that the use of temporary[283,284] screens, whilst useful to solve short-term difficulties, should be a method of last resort. Temporary screens may not be used because of operational pressures, they may be damaged or may be used as a substitute for quieter plant.

There is one reference to the use of 'white noise'[109] to mask operational noise. An innocuous continuous noise is broadcast at a relatively low level from a number of loudspeakers around the site. In effect it reduces the contrast between operational noise and the natural ambient level which can be very low. Masking noise is used in a number of open plan offices with considerable success. However, unless such an approach has been 'sold' to residents by a visit to a successful installation, there is likely to be considerable antipathy towards the idea.

In cases where BPM is being or will be used on site and the noise levels are still high, the alternatives are to:
- cause a significant nuisance,
- have planning permission refused,
- stop work,
- provide compensatory measures, such as secondary glazing to nearby houses.

5.4.4 Future Good Practice

EC directives and corresponding UK regulations[282] on noise increasingly mean that mobile plant has to meet noise limits in order to be sold in the European Community. As existing plant is scrapped, the plant pool will on average become quieter and the availability of noise emission data will improve. In the shorter term, one possibility is to create an industry-wide data base for those items of plant which have been noise-tested, and to create pressure for non-documented plant to be measured or its use to be avoided in sensitive situations.

Reversing alarms are being developed which are either inherently less disturbing or are discriminating and only operate when there is a somebody in the vicinity who should be warned. It is hoped that these more 'neighbour-friendly' alarms will become widely available.

5. Noise

5.5 RECOMMENDATIONS FOR FURTHER ADVICE AND RESEARCH

(1) Some guidance is likely to be welcomed on setting reasonable noise limits for different periods (day, evening, night). Special limits for the early morning or dawn period as well as the evening should be seriously considered. The desirable variation of these limits should be discussed with respect to:
- ambient levels,
- duration of the working or activity (1 week, 1 month, 1 year, 5 years, 50 years or more),
- whether or not there is a succession of workings,
- type of area.

Guidance is likely to follow from the WS Atkins report[267]. Whilst the consideration of acceptable noise levels requires a framework, possibly in the form of guidance from the Department, the final decision about what are desirable limits on extra noise should be site-specific and where possible compatible with environmental and planning aspirations for an area. Otherwise there will be no concerted policy for an area and its environment. Initially BS 5228 gave guidance on 'acceptable' noise levels near construction sites. This was withdrawn when the code of practice was revised because the level was being taken as a standard rather than as a guide to a reasonable upper limit. The fears of operators that low levels in quiet areas will be seen as precedents and applied more widely are understood. They should not however inhibit the setting of site-specific limits.

Work will be needed on the relationship between the duration of workings or activities and the levels of noise which cause nuisance.

Acceptable night-time levels should be examined further particularly with respect to the use of draglines and specifically the effect of intermittent clanking and squealing of chains and buckets. The noise from pumps in terms of tonality and ease or otherwise of reduction should also be considered. It may be appropriate to specify a maximum level as well as an L_{eq}. Any conclusion may have to rely upon a social survey aimed specifically at this issue; an alternative is a series of case studies based upon records of complaints, although we suspect that sufficient documentation is unlikely to be available.

(2) There is an urgent need for readily available less intrusive reversing alarms[240,267,460] which are effective, robust and cheap, and for their use to be encouraged. The intrusive non-discriminating alarms are relatively cheap and robust but their persistence may cause workers to become accustomed to them and to ignore them. Quieter alarms may be acceptable in quieter areas.

After dark, flashing lights appear to be acceptable to HSE as reversing alarms and to neighbours so long as they do not cause visual nuisance and so acoustic alarms can be switched off. Discriminating alarms which sense obstructions and alarms which issue a verbal warning are less robust and more expensive than acoustic alarms. Advice from HSE is urgently required on acceptable options for reversing alarms which are not reliant on intrusive noise[460,466].

(3) Circular 10/73[297] should be reviewed and consideration given to inclusion of noise from surface mineral sites to deter undue encroachment by development within the lifetime of long-term working such as quarries.

(4) Whilst BS 5228 could be updated to include modern equipment[430], a more practicable and useful alternative would be to create a separate industry-wide data base which could be updated on an ongoing basis. It would be helpful if such a data base were to include specific items of plant[240] discriminating between quieter and noisier models, both new and old, of equipment and indicate the availability and benefits of various levels of retro-fit. In the longer-term the effect of EC standards will simplify the problem. It will however be twenty or so years before all the relevant plant is covered by such standards and existing 'non-conforming' plant is no longer in use. Much work is needed before there is sufficient information to establish such a data base.

Clearly this is not something for the Department alone, indeed its involvement should only be as a catalyst. Liaison with the Construction Industry Research and Information Association with whom there are overlapping interests, the minerals and construction industry in general and the manufacturers of plant and equipment would be beneficial.

(5) Encouragement should be given to any attempts to produce quieter empty vehicles, in particular to reduce body-slap (see also Chapter 3).

5. Noise

SUMMARY OF GOOD PRACTICE – NOISE

Good Practice for Mineral Planning Authorities

- consider the ambient noise, planning policies and the duration of the noise; discuss any limits and monitoring with the local Environmental Health Officer.
- consider the need to agree or specify planning conditions relating to:
    - noise limits at sensitive houses, etc, for various periods of the day,
    - the provision of monitoring equipment,
    - limits on hours of operation,
    - noise control measures, eg gradual start-up of plant,
    - adherence to a code of practice,
  or in the last resort, usually for particular activities:
    - noise emission from plant temporarily working close to houses,
    - types of plant and/or number of items in use simultaneously.

Good Practice for Operators

- discuss noise in advance with the MPA and demonstrate in their application that proposed conditions can be met,
- plan ahead and make sure that:
    - noise is a factor in the layout, and the nature and sequence of working,
    - work at night near sensitive areas is avoided where possible,
    - screening is part of the design, eg by bunds and working face,
    - the quieter of the methods or plant available is chosen,
    - especial care is taken with reversing alarms,
    - haul-roads are screened and without severe gradients,
- ensure that management has the will to run the site as quietly as possible,
- check the noise characteristics of plant before use and periodically thereafter, where appropriate retro-fit noisy plant, ensure good operation and maintenance,
- make no unnecessary noise and reduce noise emissions, eg:
    - minimise height which material drops from lorries or plant,
    - minimise distance between loading and emptying dragline buckets,
    - reduce clanging of dragline buckets & chains by careful operation,
    - use rubber linings in chutes, dumpers, trucks, transfer points,
    - clad plant and ensure that the cladding is kept free of holes,
    - start items of plant one by one, possibly behind mounds,
    - switch-off equipment when not in use, avoid unnecessary revving of engines,
    - keep noise control hoods closed when machines are in use,
    - keep lorry tailgates closed where possible,
- as a last resort, reduce the propagation of noise, by the use of:
    - temporary bunds,
    - portable screens.

CHAPTER 6

AIR POLLUTION – DUST, ODOUR

6.1 GENERAL

Air pollution is defined here as dust or odour. The main potential effect of odour is to cause nuisance to neighbours.

Dust is considered to be any solid matter emanating from a surface mineral working, or from vehicles serving it, which is borne by the air and can range in size from the smallest individually invisible smoke particle up to about 2 mm. It can be emitted from a stack as a plume or it can be picked up by the wind from the ground, the surface of a road or a stockpile. Depending upon their chemical composition, the particles can be chemically active, eg limestone, or effectively inert, eg sand. Their colour varies from black, eg coal, through brown to white, eg cement or chalk. The finest particles will be respirable.

The main potential effects of dust are:
- visual; dust plumes, reduced visibility, coating and soiling of surfaces leading to annoyance, loss of amenity, the need to clean surfaces,
- physical and/or chemical contamination and corrosion[393] of artefacts leading to:
    - a need for cleaning,
    - mechanical or electrical faults, eg with computers, electro-mechanical devices,
    - abrasion of moving parts,
    - soiling of finished products, spoilt paint or polish finishes,
    - contamination of laboratory, quality control, standards room and medical facilities,
- coating of vegetation and contamination of soils leading to changes in growth rates of vegetation and possibly reduced value of agricultural products,
- health effects due to inhalation, eg asthma, or irritation of the eyes; metalliferous mining is not included in this study so toxic effects do not have to be considered.

There are a number of mechanisms which produce dust[238]. It can occur naturally, as in sand, or it can be produced by fracturing larger particles or aggregations. Small particles can be picked up by the wind from exposed soil surfaces and carried long distances in suspension. Wind over a stockpile or through a stream of falling material, as off a conveyor[301], can also

6. Dust

remove fine particles. A falling stream of material can itself create an air current with a similar effect.

There are three main types of dust source:
- **point**, eg drilling of hard rock, blasting, loading, tipping, earthmoving plant, draglines, dozers, chutes, crushers, screens, exhausts from dust control systems, dryer chimneys, unsheeted trucks, conveyor transfer points,
- **line**, eg well-defined haul-roads, open conveyors,
- **area or dispersed**, eg top/sub-soil stripping and dumping when dry and friable, quarry floors, unsurfaced haul-roads, waste dumps, stock piles, sand faces, spillage.

Some measurement data[214] and some methods of prediction[181,182,202,235] are available for dust generation and propagation. Some emission predictions are possible[394] based upon measured rates of dust production by particular operations and vehicle movements. Predictions of dispersion are normally based upon a Gaussian plume model (for point and line sources). Deposition rates depend upon particle size. Overall, the resulting predictions are likely to be qualitative and relative rather than precise; many sources and the extent of propagation vary considerably with the weather, especially rainfall and wind.

6.2 PROBLEMS

6.2.1 Dust

Dust from workings is often a significant nuisance[47,237,286,429,455]. In one case sand and gravel workings caused problems at 350 m. Premises are said to be 'vulnerable' within about 500 m[393], but the distance varies with local topography and prevailing winds. Dust levels upwind of workings are likely to represent the ambient level, downwind of workings the levels can be 2-3 times higher[378]. At opencast sites, dust is said to be the second major source of irritation, although this is usually only at a few sites and at specific periods of their lives[393].

In residential areas dust can soil washing, prams, gardens, windows, paintwork, cars and enter houses. Soiled/discoloured vegetation (near quarries or along roadside verges) and buildings[226,286], dust plumes from plant, dust clouds trailed by vehicles[439] and even dusty plant are seen as 'eyesores'. Road signs can be obscured. The general dinginess can affect property values[393]. The possibility of damage due to corrosion has been suggested[262] but not demonstrated.

In scenic areas[353] especially on dry windy days, significant visual intrusion is caused by dust coating vegetation and affecting the views over many kilometres. Overall there is a reduction in the amenity value of footpaths, viewpoints and the area in general.

The weather makes a vast difference to dust emissions and to its spread. Soil working[237] during dry weather creates dust which is readily removed by wind or even rain. However dust deposited during periods of light rain causes encrustation which can curtail plant growth and fruit-set.

Some soiling of agricultural produce[288] and pasture has been suggested anecdotally. The former could affect the commercial value of the produce for human consumption; the latter is said to have affected the uptake of nourishment by cattle. It appears almost certain that some operators have paid compensation for such episodes but the scientific 'justification' is unknown. On the other hand, lime dust could be beneficial to cattle pasture[466].

The effects on vegetation[202] appear to be limited and largely unexplored[461]; at worst, significant effects occur infrequently. There is some evidence of adverse effects on growth and diversity of woodlands adjacent to quarries[248], although apart from dust, these effects may relate to changes in drainage or groundwater and wind exposure.

Four species of tree were studied in an area near to a limestone process plant and in a control area. In the affected area three species declined but the fourth flourished presumably because it is lime-tolerant and because of reduced competition[312]. The adverse effects were said to be due to a crust on the leaves reducing photosynthesis, inhibiting growth of new photosynthetic tissue, destroying leaf tissue and causing premature leaf fall. The overall effect was to lead to a different dominant species[313]. Elsewhere, limestone dust has been found to affect limestone heathland[4]. Cement dust has been found to affect some species of agricultural crop[381]. On the other hand a limestone processing plant in a disused quarry designated as a SSSI has operated without any apparent deleterious effects[286]. The pH levels on tree bark are affected by alkaline dust up to 5 km downwind from a quarry and cement works[162]. This has caused species substitution and a greatly increased species diversity of epiphytic lichens. It is not clear whether this is to be regarded as good or bad!

Some observation of the yellowing of grass near a slate quarry has been reported[262].

There is some suggestion that vegetation and cattle near to fletton brickworks have augmented levels of fluoride and may suffer from fluorosis[434].

It has been suggested that the eyes of blind people may be unusually sensitive to dust[202]. It has also been suggested that dust causes health problems off-site[29] but this has been disputed[89]. There are no credible data on health effects at levels appropriate to residential areas[458] and it is unlikely that respirable dust is a problem[458], however larger non-respirable particles could be an irritant to the eyes, nose and throat. Further consideration of this subject is outside the remit of this study.

In addition to problems off-site, on-site dust may[417]:
- reduce visibility and safety,
- cause wear in engines, bearings,

6. Dust

- lead to higher maintenance costs due to increased frequency in changes of oil, filters,
- increase tyre penetration on dusty roads and fuel consumption of trucks, etc,
- create unpleasant and potentially unhealthy conditions for workers,
- cause damage to revegetation schemes.

6.2.2 Odour

Odour nuisance occurs occasionally. Sometimes the burning of waste oil in process plant and poor housekeeping in the use of volatile materials in roadstone coating plant lead to significant odour episodes, the smell of unburnt fuel and smoke.

The clay used to produce fletton bricks is high in organic material and the brick making process leads to smell of mercaptans[434]. This can cause complaints over a fairly wide area. Some work has been done to try to burn-off the smell[433] but without great success.

No measurable physical criterion exists for odour and there is little literature on the subject relevant to surface mineral workings. Beyond noting that the problem exists there is little that can be said except in specific cases. This report will not return to the subject.

6.3 ACCEPTABLE LEVELS AND MONITORING

6.3.1 Acceptable Dust Levels

Dust pollution, as all air pollution borne on the wind, can vary rapidly; typically daily peaks may be 2-5 times the monthly average[393]. Unfortunately there are problems with the measurement of dust, in establishing reliable criteria and setting credible limits for it[393]. There are few data on what particle sizes cause dustiness[181]. Occasionally dust levels have been specified in planning conditions[426] in spite of the uncertainties involved but not as far as we know in relation to surface mineral workings.

Human reaction to the overall deposition of dust can relate to the rate of deposition, ie how quickly things become dusty and the degree of dustiness, or to the level of dustiness by contrast with other cleaner areas, eg when a surface is partially cleaned by finger-marks[301]. Significant nuisance is likely when the dust coverage of surfaces, eg windowsills or cars, is visible in contrast to adjacent clean areas, especially when it happens regularly[393]. Severe nuisance occurs when the dust is perceptible without a clean reference surface.

It has been suggested[202,236] that up to 0.17-0.5% effective area coverage of a surface per day is the most that can be tolerated; 0.5% represents possible complaints and a marginal nuisance, whereas 5% represents serious complaints and would be a severe nuisance[300,390].

6. Dust

Typical ambient levels are 0.3–0.4%/d in an urban area, 0.01–0.5%/d in a rural area and 0.8–1.0%/d in an industrial area.

The other main criterion for nuisance[202,300,393] due to non-toxic dust is a deposition rate of 200 mg/m^2/d averaged over a month or 80 mg/m^2/d for black coal dust. Above these levels the need for cleaning becomes excessive. The criteria themselves are relatively unsophisticated compared to those for noise. They should be treated with caution and more experience is needed of their application.

Other criteria are based upon total suspended particulates (TSP)[393]; the US Environmental Protection Agency proposed standard is an annual average of 75 μg/m^3 which is roughly equivalent to 50 mg/m^2/d.

Acceptable levels can be related to the ambient[393], eg 2–3 times the background deposition rate, and a criterion based upon the number of times daily levels exceed this 'acceptable' level. Ambient levels are typically 10–50, 30–80 and 80–160 mg/m^2/d in rural, suburban, and town centre or industrial areas respectively. Although other sources give 65, more when harrowing fields, to 20–100[429] in rural locations, 90 for suburban areas and 160 for town centres and 130 mg/m^2/d for commercial areas.

Using any of the above criteria it may still be necessary to examine the collected dust, eg by microscope, to be sure that the dust has come from the site and not elsewhere.

There are also less precise criteria, eg that the process plant should be substantially free from visible emissions[422]. However for mineral process plant prescribed under Section 2(1) of the Environmental Protection Act 1990 (EPA) for control by Local Authorities, specific emission limits are given in guidance provided by the Secretary of State[491,492].

6.3.2 Monitoring

Measurement[174,202,303] of dust from a particular source is usually either imprecise or expensive. Total deposition gauges collect dust deposited in them which can then be weighed[394].

Directional gauges BS 1747[307] are particularly poor at measuring quantity and are not very precise in terms of direction[308]. An automatic directional dust sampler with dual heads[291] controlled and activated by wind direction and speed is likely to be more useful in identifying and quantifying sources of fugitive dust. Another simpler improved deposit gauge has been put forward[320].

The criteria available for acceptable rates of dust deposition or coverage relate to dust from all sources. As with noise, it is necessary to discriminate between sources and this sometimes requires human intervention. The automatic directional sampler referred to above may be able to discriminate sufficiently to provide a reliable quantification of dust deposition originating from a site.

6. Dust

The sticky pad method[389,390,394] is very cheap for measuring rate of effective area coverage and can be successfully used on a DIY basis. Many gauges can therefore be used to gain an understanding of where the dust is coming from and the variability of its effects in a neighbourhood. The method relies upon a sticky pad collecting dust over a given period; changes in its reflectivity compared to an unexposed control surface are measured by a readily available standard meter. Conversion curves are available to translate reflectivity into deposition rates but these should be treated with caution[456].

High volume samplers which draw air through a filter will measure the TSP accurately[394] but not rates of dust deposition. A dust gauge which measures TSP automatically and which can transmit the results over the telephone is also available[421].

6.4 GOOD PRACTICE

6.4.1 Sources of Dust

The prime sources of dust are[47,237,286,393,429,455]:– vehicles (including sweepers) on dry unsurfaced haul-roads, overburden heaps, open storage, loading of dry materials into lorries and perhaps particularly processing plant[286] especially for crushed rock or sand and gravel and when there are operational or maintenance problems. Most are worse in dry weather. Processing plant sometimes emits dust at a rate five times that of the rest of the site[214]. There is a reluctance to collect dust spillages and to dispose of them 'safely'.

The various types of dust source require different approaches to control; in particular, area/dispersed sources are difficult to contain.

Technically there appear to be a sufficient number of practical options[7,24,32,33,39,47,50,68,74,76,79,113,116,177,214,215,234,239,240,262,301,338,382,383,393,396,417,422]. The guidance referred to at the end of Section 6.3.1 relating to the EPA[491,492] contains specific detail on the type and degree of control that is expected (BATNEEC see Section 2.2.1.3) for quarry processes, including those for roadstone and coal. Essentially there are four ways of reducing dust, although the distinction between them is not always clear cut:

- foreseeing problems and avoiding them,
- preventing its escape into the atmosphere,
- recapturing it once it is in the air,
- reducing its spread once it is airborne.

As with most pollutants, to avoid creating it is easier than subsequent control.

6.4.2 Planning to Avoid Problems

To avoid or minimise the creation of dust and/or problems one can:
- use conveyors[397] in preference to haul-roads,
- locate haul-roads, tips and stockpiles away from sensitive neighbours taking account of prevailing wind directions[393],
- design haul-roads with dust in mind, by minimising length and gradients as far as is mutually compatible, surfacing roads where possible, limiting vehicle speeds,
- keep surge piles to a minimum; site emergency piles in sheltered areas[422],
- design tips and mounds so their shape, during construction and use, minimises the accentuation of surface wind speeds and hence the pick-up of dust[234], ie use gentle slopes, avoid sharp changes of shape; compact and bind the surfaces,
- use permeable fences to reduce wind speeds to avoid the pick-up of dust or to help it to settle out if it is airborne[234,262,393],
- plan to have 'dust-sensitive zones' within the site boundary[383]. In these zones, which may be near to sensitive properties, there would be no stripping of top/sub-soil in dry weather and no storage at any time,
- use chippings[289] instead of dust for stemming blast holes,
- shatter rock to the least extent necessary,
- use crushing and screening plant within its capacity[455],
- minimise the height of fall of material,
- grade, vegetate[477] or stabilise, eg with polymers, long-term overburden mounds, waste heaps and tailings.

6.4.3 Controlling the Escape of Dust

To control dust (mainly of point or line sources) containment[370], including that of stockpiles[492], can be used to prevent spillage and emissions. There is the disadvantage that containment may conflict with access[455] and hatches may be left open. Ideally, the inside air should be at a negative pressure relative to the outside so that any air flow is inwards[301].

The following controls have been suggested:
- closed or sheeted vehicles for internal transfer of dusty material[422,492],
- the enclosure of conveyors, chutes, process plant, and avoiding holes in screening and crusher houses; have self-closing doors[422,492],
- the use of water sprays or preferably mists, foam[113] or microfoam[48] to minimise the wetting of the material to be processed,
- an internal dust removal system for the plant[208],
- outlets fitted with cyclones, wet-scrubbers, electrostatic precipitators (rarely justified), filters (bag, etc),
- the pelletisation of dust or other means of securing it prior to disposal[45],
- good maintenance.

6. Dust

The disposal of collected dust, whether dry or as a slurry, may be an environmental problem[458] which needs to be addressed. In one case[353] quarry dust from bag filters was passed to a silo and then taken by tanker to an asphalt plant for use as a filler.

Partial enclosure can also help, eg one could:
- fit dust extractors, filters and collectors on drilling rigs (wet with foam or dry)[177]; wet-drilling methods are an alternative but may affect the charging of the hole,
- use mats[262] when blasting (designed for flyrock),
- create dust only in sheltered areas[234],
- use trees or shrubs[393], wind-breaks/netting screens/semi-permeable fences[234] (those that can resist high winds will be expensive) on[393] top of bunds, around stockpiles, along haul-roads, across tips, around loading shovels,
- fit wind-boards to conveyors[369], hoods at transfer points, limit height of fall[301]; but be aware that telescopic or other more complete hoods may entrain draughts.

The main possibilities for sources which cannot be enclosed are to:
- compact, grade, surface and maintain haul-roads,
- reduce speeds and limit the movement of vehicles, use upswept exhausts, blow air forward through radiators; enclosure of wheels does not seem to help[417],
- vegetate with quick growing plants or spray areas which would otherwise remain exposed for three months or more with binders,
- limit spillage[301],
- facilitate the removal of spillage by the use of hard surfaces, eg concrete, where spillage is likely, especially in loading areas, under and around stockpiles,
- sweep/vacuum haul-roads and other dusty surfaces (but choose equipment carefully[380] depending upon the nature of the surface, ie rough, smooth, wet or dry, otherwise the cleaning equipment itself may raise dust),
- shake-off dirt from vehicles by requiring them to be driven over a grid,
- provide wheel, mud flaps and lower body washing facilities prior to the use of public roads,
- provide a length of surfaced road (after washers) before the exit from the site,
- use closed or sheeted vehicles when carrying dry materials or a material whose surface may become dry in transit; this can apply to 'empty' vehicles as well because of the residue left after tipping. Semi-permanent proprietary sheeting is available[301] which is reported to be superior to the tied-on tarpaulin and allowing easy loading. When lorries are dedicated to one material it may be worth equipping them with a permanent metal or plastic cover,
- use water and/or binders on finished earthworks, unsurfaced haul-roads (see discussion below for fuller details).

A more extensive discussion of the cleaning and sheeting of road vehicles is to be found in Sections 3.4.3 and 3.4.4.

6.4.4 The Use of Water to Suppress Dust

Water is the most readily available means of suppressing dust. The use of additives to improve its performance is reviewed in the next section. It is possible to:
- use water bowsers, sprays or vapour masts (the latter can also remove dust from the air) on haul-roads and hardstandings,
- control sprays and vapour masts by using sensors[369], eg photoelectric cells to detect material on conveyors or the opening of hopper doors; clocks and anemometers[301] are other possibilities,
- spray stockpiles, other mounds of material and bare ground with water, solutions or suspensions; on penetration a cohesive protective layer will be formed. Where possible material should be treated prior to stocking or piling and then wet again before being disturbed in dry weather[422]. If the protective layer is disturbed the application will need to be repeated. The applications, via sprinklers or sprays should be activated by wind-speed and rainfall[301].

These measures generally reduce dust to less than 10% compared to that from an untreated surface and last about 50 days[234]; water is said[239] to be up to 96% effective but only when the roads are still thoroughly wet or the 'crust' on materials is unbroken. The results can be slightly better for individual sources, eg crushers, conveyor transfer points.

Sprays[214] need inspection and maintenance to avoid blockages but are much cheaper than containment. The use of bowsers, sprays or vapour masts need an adequate and reliable source of water although mist sprays require minimal water. The supply and use of large quantities of water may be difficult to maintain in dry weather when it is most likely to be needed, although sump water can be used if sufficiently free of solids[396]. Fine sprays with wetting agents and foam[113] reduce the need for water and wet fine dust more effectively than bowsers or coarse sprays. In cold weather water may freeze and not be available when needed[455]; ice may also be a danger on roads.

When water is used to control dust from processing plant, the end product may be undesirably wet or, if dried, may be coated with a film of dust. In bad cases, processing of wet stone may lead to clogged screens or spillage at conveyor pulleys. This may require machines to be stopped and guards removed to clear the problem, which can lead to accidents unless the procedure is properly supervised[458].

In both contexts, ie water usage/supply and process quality, the use of fine sprays, vapour masts and additives offer potential advantages compared to the use of plain water by bowsers and coarse sprays.

6.4.5 Use of Binders and Additives

A survey[24] of a number of stone processing plants indicated that whereas about 90% of quarries used water to suppress dust, only 37% used those chemicals necessary to wet fine

6. Dust

dust. Anecdotally it is suggested that, whilst most dusty workings use water sprays, only a minority use additives or dust extractors.

Water, whilst effective when it is first applied does not have a long lasting benefit[417]. For semi-permanent haul-roads a binder which retains moisture is claimed to be helpful[113,214,417]. Salt brines have been used for a long time and are said to be effective but may need to be replaced after heavy rain and may cause problems with vegetation, groundwater or surface water. Calcium chloride is used by spreading evenly on road surfaces in dry weather[458]. Magnesium chloride has been found to be 95% effective on sand and gravel roads[417]. Waste oils and refined oils are also effective but are potentially environmentally damaging. They should not be used in river valleys or at sand and gravel sites where the groundwater level is high[460]. No binders should be used without consultation with the NRA. In some cases the use of binders may not be practical where haul-roads are regularly scraped and regraded.

Many proprietary and other materials have been developed and are available to extend the effective life of water as a binder. They are based upon adhesives and chemicals with film-forming properties; additives include latex, asphalt, waxes and polymers to counteract drying. Most binders are applied superficially but some involve scarifying, regrading and compacting; it is claimed that the additional material costs are offset by lower labour costs due to less frequent applications. However they do tend to be washed away by heavy rain and overwatering[184]. It may be necessary (and usually wise) to test additives and binders on site[33].

Materials for use on soil have been categorised according to function[388]. 'Binders' protect bare surfaces from erosion until vegetated and include emulsified asphalt, PVA or PVP. 'Surface protection composites' stop erosion by providing a physical cover and may be chopped straw or polyethylene mesh; they are permeable to water and plants grow through them. 'Mulches', such as wood pulp, also improve the soil structure. 'Water retention aids', such as 'Aquagel' retain 30 times their volume of water without an effect on pH, calcium chloride is hygroscopic but can change pH; it is acceptable on roads but questionable for areas to be vegetated. The same report[388] reviews the materials commercially available and gives an indication of any known side effects. Lignin sulphonate has been found to be 90% effective on tailings[417]; neutralised with lime it helped revegetation.

All of the above practices are claimed by their protagonists to be economic and effective but more data are needed, especially about the effectiveness of additives, the treatment of haul-roads and the use of foam.

Little practical information is available[184,263] and most users appear to have conducted their own site trials before finding an appropriate way of using proprietary materials. In some cases, it is possible to make up a site-specific formulation more cheaply by substituting a DIY equivalent to proprietary materials but NRA should be consulted. The results of this experience are generally unpublished[428].

During the 1980s 'residual' dust suppressants[113] were developed which remain effective through several successive stages of materials handling and processing. They can be used

6. Dust

on active piles; each time a new surface is exposed the suppressant is activated to form a fresh coating or to cause the agglomeration of fines. Residual suppressants are generally combinations of wetting agents or foams with latex, lignosulfonate or other film-forming adhesive compounds.

6.4.6 Removal of Dust from the Atmosphere

To control dust by removing it from the atmosphere one can:
- use fine water sprays/mists[369], mobile vapour masts[68], additives[215] (surfactants or wetting agents),
- use trees or shrubs[393] around the site to filter out dust or to cause the wind velocity to slow and dust to drop out of suspension; bunding is said not to be very effective.

BCOE have carried out trials of vapour masts[358] and the benefits have led to their increased use. The efficiency of trees and shrubs as filters will depend upon the species and whether they are evergreen or deciduous. Compatibility with the ecology and landscape of the area would need to be considered.

If all else fails it may be that an operator will have to consider temporarily stopping some or all dusty activities. This may arise during a period of particularly dry and windy conditions in order to protect neighbours. Casting of material, the tipping of overburden or waste, above ground haulage or the grading of overburden[393] may have to be curtailed when the wind is towards a sensitive area. As an aid wind-socks can be used on mounds[453]. BCOE empower their Site Engineers[351,358,474] to stop their contractors from working in the event of adverse conditions causing dust to be a nuisance to neighbours. Some planning conditions also require this[359].

The handling of soil will require careful judgement[383]. At one extreme it is necessary to avoid handling in a wet and coagulated condition because it is difficult to handle and may damage the soil structure. At the other extreme, in a dry conditions too much dust may be created especially if it is windy at the same time.

6.4.7 Future Good Practice

Compared with noise the subject is difficult to quantify. Although the technology of control seems to be well advanced it is unquantified, under-documented and probably under-used. When more data and practical experience is available about the use of environmentally acceptable binders, wetting agents and other additives, we expect the control of dust to be significantly less difficult and the results more effective.

At the present time it is not generally possible to rely upon performance criteria because of the uncertainties in monitoring. Neither is it possible to rely upon specifying control measures which are both reasonable in terms of cost and effective in avoiding problems. The

6. Dust

application of BPM is equally uncertain because there is no clear understanding, without a code of practice, of what it implies. A code of practice or a reliable criterion which could be readily monitored would improve the situation.

In the medium-term, one suggestion[393] which we consider deserves consideration, combines requirements for monitoring with specific control measures. A numerical criterion is set as a warning significantly below the level at which nuisance might be expected so that its value is not critical. If the warning level is exceeded, the application of the agreed control measures is triggered. Thus the control measures are only used if they prove to be necessary.

More monitoring of complaints and of dust levels in whatever ways are available will help to improve the knowledge and understanding of nuisance levels.

6.5 RECOMMENDATIONS FOR FURTHER ADVICE AND RESEARCH

(1) A review of dust control methods should be commissioned with a view to forming the basis of a code of practice. This should include the design of stockpiles, means of reducing surface wind speeds, the effect of baffle banks and wind flow on the pick-up of dust, as well as the costs, use and effectiveness of additives, binders, foam and residual suppressants. Guidance could be issued as part of a code of practice rather like that produced for the materials handling industry[301].

(2) There is a need for dust nuisance criteria and more reliable means of dust measurement and prediction. Until these are available, it is difficult for planning authorities to set performance standards and leave the operator to organise working procedures in the most economical way. Research should be commissioned to confirm or develop criteria and to recommend acceptable levels and methods of measurement. A social survey of public response to various levels/rates of dust deposition should be included in the work.

(3) The suggestion of a hybrid approach to planning control with a warning or 'trigger' level signalling the need for the implementation of specific agreed control measures needs to be examined. If appropriate, advice should be provided to MPAs. As well as offering a way forward, this approach could form a useful basis for improving the state of the art in relation to criteria and monitoring.

SUMMARY OF GOOD PRACTICE – DUST

Good Practice for Mineral Planning Authorities

- liaise with District Councils if the process is prescribed for their control under the EPA,
- consider the need to agree or specify planning conditions relating to the:
    - layout of the site, design of stockpiles,
    - containment of conveyors and processing plant and dust collection equipment,
    - use of bowsers, sprays and vapour masts on haul-roads, stockpiles, transfer points,
    - design of material-handling systems, drop heights, wind guards, loading points,
    - use of binders on haul-roads and stockpiles (after consulting NRA).
    - limiting levels of dust measured in a specific way; provision of monitoring facilities.

Good Practice for Operators

- minimise the creation of dust by planning and design, eg:
    - use conveyors rather than haul-roads,
    - locate haul-roads, tips and stockpiles away and down-wind from neighbours,
    - create 'sensitive zones' within which activities are limited,
    - layout and construct stockpiles, tips and mounds to minimise dust creation; use gentle slopes and avoid sharp changes of shape,
    - use crushing and screening plant within its design capacity,
    - minimise the height of fall of material,
    - use appropriate chippings for stemming,
- control the escape of dust, eg:
    - enclose conveyors, chutes, process plant, stockpiles,
    - provide of a dust removal system for the plant,
    - use sprays, mists, microfoam or foam,
    - fit outlets with cyclones, wet-scrubbers, filters,
    - pelletise dust or otherwise make it secure prior to disposal,
    - insist on good maintenance,
- minimise dust pick-up by wind, eg:
    - compact, grade, surface and maintain haul-roads,
    - fit dust extractors, filters and collectors on drilling rigs,
    - use mats when blasting,
    - restrict dust-making activities to sheltered areas,
    - use wind-breaks/netting screens/semi-permeable fences,
    - limit drop of falling material,
    - fit wind-boards/hoods to conveyors/transfer points,
    - reduce speeds and limit movement of vehicles, use upswept exhausts,
    - use water bowsers, sprays or vapour masts,
    - spray exposed surfaces, eg unsurfaced haul-roads, stockpiles, with binders (consult NRA),
    - vegetate exposed surfaces, eg overburden mounds, with quick growing plants,
    - limit spillage; facilitate the removal by the use of hard surfaces,
    - sweep/vacuum haul-roads and other dusty surfaces,
    - shake-off dirt from vehicles, provide vehicle washing facilities,
    - provide a surfaced road between washing facilities and site exit,
    - use closed or sheeted vehicles carrying dry material,
- remove dust from the atmosphere, eg:
    - use fine water sprays/mists, with or without additives,
    - use trees or shrubs around the site,
- stop the activity or operations, if the creation of dust cannot be avoided.

CHAPTER 7

LANDSCAPE

7.1 GENERAL

The effects of surface mineral working can be:
- to destroy some of the existing landscape, eg a hill. This is outside the remit of this study, but whatever is done by way of amelioration during or after operations cannot alter this loss, restoration can only provide a new landscape albeit similar to the original,
- to introduce a feature into the landscape which may be alien to it and create a visual intrusion, eg, a quarry face, an overburden mound or machinery,
- to screen from view some of the landscape that is otherwise unaffected.

7.2 PROBLEMS

Both the literature and our discussions indicate that landscape changes and visual intrusion are major planning problems but do not give rise to as many complaints as might be expected during operation. The main damage is often inherent in the planning permission and only partially ameliorated by subsequent 'cosmetic' treatment. The absence of a mature landscape[351] for decades, even with subsequent full restoration, may be the major loss. Complaints may be seen as pointless, be transferred to more tangible matters, eg traffic, or reserved to reappear later as objections to future proposals.

Complaints are more likely to arise from abrupt changes, such as the erection of a noise fence or bund, or the arrival of process plant, than from the slow changes brought about by excavation. A freshly exposed rock face is noticeable[177] by contrast with its surroundings and a break in the skyline is particularly obvious. The size of a mine/quarry is not of itself necessarily a problem unless it is out of scale with its surroundings. Vehicles on access roads[201] and especially processing plant can be visually intrusive. Lights in workings frequently give rise to complaints[455] due to their adverse effects on amenity, the glare they cause and the distraction and danger to traffic.

Complaints are received about landscaping measures themselves. Mounds and screen-planting are not always desirable features. Height limits sometimes result in overburden mounds having flat tops which are unsightly and result in poor run-off. Although screening mounds may generally be better than the mineral workings, they may be seen as alien

Environmental Effects of Surface Mineral Workings

7. Landscape

features in the landscape, especially if obviously man-made[455]. Screening mounds and noise bunds are sometimes regarded as unsightly in themselves and badly contoured banks may appear worse than the quarry[177]. Some residents would rather have some view than nothing but the face of a screening mound[247]; a mound may also interfere with TV reception. In one case residents preferred not to have a noise fence on top of a mound.

Landscape conditions attached to industrial planning permissions have often been ignored[228] and those relating to surface mineral workings are no exception; preplanting conditions are neither always implemented nor enforced. Planting single rows of trees, particularly in locations where trees do not normally grow due to an inappropriate climate, is likely to be no more than an ineffective gesture; even if successful, the result would be out of character[286]. Planting is not always well maintained[353], eg growth may be choked by weeds and dead trees not replaced.

The entrance to a working is often the most noticeable part, especially where the process plant is adjacent to it. Plant which is derelict, dilapidated, untidy or inappropriately sited[15] is a significant source of local complaint. Company colours can be as unwelcome and intrusive as rusty or dirty colouring.

7.3 ACCEPTABLE MEASURES AND MONITORING

Clearly there are few objective criteria and standards which can be specified and monitored. Of those that can, most are fairly readily monitored, ie:
- the timing of planting,
- the limits of the quarry face,
- the programme of working and progressive restoration,
- the siting and dimension of screening mounds,
- the siting and shape of plant and of overburden mounds.

It is more difficult to monitor the quality and success of planting and vegetation.

Opinions vary about whether mounds should be naturally revegetated or artificially sown. One of the problems with natural revegetation is that the initial growth is by the invasive species, ie weeds, and it takes longer for the balance of the new growth to match that of the surrounding vegetation.

It is necessary to anticipate problems by monitoring the preparation prior to planting, eg the soil and plant studies and trials. As we have seen above, often the problem has been noticed when it is too late to remedy it.

7. Landscape

7.4 GOOD PRACTICE

The basic options are to:
- avoid problems by planning, especially to incorporate progressive restoration,
- improve the appearance of the quarry and the processing plant,
- conceal workings by screening.

However, these are not separate activities and often need to be considered together. In some cases little can be done by way of detailed operational control after working starts.

7.4.1 Planning

Avoiding problems by planning means that it is necessary to take a positive and integrated approach to the consideration of landscape together with the planning of the mineral workings, pattern/scheduling of excavations, siting of plant, haul-roads, etc. We believe that computer-aided-design can be of great assistance in this approach. Many of these operational features need to be considered before the granting of planning permission.

In addition to considering the limits of excavation, the early planning should test the various options for the sequences and directions in which the site can be worked against visual criteria before making a final choice[285]. It will often be advantageous to work towards a sensitive area both for visual and noise reasons because of the protection offered by the working face, ie to use the existing landscape to screen the workings.

7.4.2 Progressive Restoration

In the case of many sand and gravel workings[6,460] and in some opencast coal workings, it is possible to work the site in a series of sectors that are progressively restored[69,432] so that the working area at any one time is relatively small and the period the land is out of 'cultivation' is much less than the total period of working. In some cases temporary 'restoration' may be appropriate.

Such an approach is also possible to some extent with hard rock quarries. At least one proposal[285] exists (now permitted) where faces will be subject to restoration blasting during the operational life of the quarry. The intention is to make the final quarry faces form a landscape sequence which 'mimics' surrounding dalesides and on which a self-sustaining ecosystem may be established[494]. This approach is being developed and assessed in a current DOE research project. Rock faces can be prepared[353] by creating ledges and scree which are treated with slurry and sprayed to assist revegetation. Waste material can be used as it is produced to contour quarry walls to blend in with the surrounding landscape.

Progressive restoration commands widespread support, eg BACMI[339], SAGA[343], BCOE[351,352] and other operators[364]. Many MPAs include it in their Structure Plans; one indicates, perhaps

Environmental Effects of Surface Mineral Workings

7. Landscape

a little hopefully, that permissions will be more likely where the rate of restoration equals the rate of working[400].

Progressive restoration may rely upon a licence to import waste, especially where the restoration is to the original ground level, eg for agricultural purposes.

One operator has a 'farmer' who looks after the agricultural use of land that is not yet being worked, tends preplanting, tends the revegetation of restored areas and the reintroduction of agriculture.

7.4.3 Siting of Overburden, Waste, Haul-Roads and Plant

Where waste from the mineral working cannot be immediately used to restore worked areas, its siting and treatment must be considered as part of the initial planning. If it cannot be hidden from view, it should be landscaped and vegetated at as early a stage as possible.

The possibility of transferring overburden within a site, eg to form part of progressive restoration, or in very sensitive areas[112] from one site to another, could be considered to avoid the need for large obtrusive mounds. The economic and environmental costs of transport will make this measure rare; two sites would need to be very close together.

Among the more obvious of other desirable landscape features are:
- a curved access road into the quarry,
- no breaks in the skyline,
- varying the landscaping to suit the various stages of mining,
- wet working which may be more visually acceptable than dry working[364] which exposes an uneven floor,
- haul-roads below the rim of the quarry[285],
- the use of conveyors, these can be less conspicuous than haul-roads[384],
- the use of low profile processing plant[6.34], eg only 50% of the height of conventional plant[372].

It is stated[339] that "fewer and fewer facilities can be seen on site these days. The plant is screened, normally by bund walls. These act as noise barriers as well". Whilst this is an improvement, it is also suggested[372] that a preference for 'low-profile' plant be included in planning policies to encourage further improvements by its use and further development of the concept. In one case where normal crushing and screening plant would have been 14.5 m high, the low profile alternative installed was 9 m high.

Once a quarry is established, plant which was initially adjacent to a public road, can be moved onto the quarry floor[286]. Reasons can always be found to delay such a move[354], but it is one of the actions most likely to improve the image of a working.

7. Landscape

7.4.4 Making the Best of the Site

Making the best of the appearance of workings could involve:
- avoiding obviously man-made banks[455], eg giving overburden mounds rounded profiles in cross-section and undulating in long-section; such contours would also help run-off and avoid ponding[177]. Mounds, banks and bunds in a rugged landscape may need to be designed differently,
- providing low mounds along the verges of hillside access roads to screen the scar,
- using low-cost swards to green temporary mounds, these will also help stabilise the slopes to minimise erosion and dust; maintenance will be needed to control weeds by mowing or use as pasture; it is fair to say that ecological opinion is divided as to whether to encourage natural vegetation or to seed[455], hydro-seeding could be used more frequently[353,357],
- spraying exposed faces with colour, as some operators have done, to reduce their obtrusiveness[109],
- reinforcement planting[383] to fill gaps in hedgerows,
- having tidy, well maintained and colour coordinated plant, well designed and landscaped entrances and visitors areas (see also Section 12.3 'Operator's Image').

It is suggested[67] that there is an overuse of screening mounds, which involve expensive double-handling of soil/overburden and that it is cheaper to pre-plant trees using good arboricultural practices to increase the success rate[493], eg selection of species, planting methods, appropriate fertilisers and maintenance. Screening mounds are artificial and fencing with tree planting may be better[358]. Off-site planting/screening might also be less obtrusive than screening mounds in some cases. This raises the difficulty of landownership[177] but it is sometimes possible to negotiate leave to plant or to acquire the land. Another possibility is to let hedgerows grow taller[384].

To make the best of planting and vegetation, whether for screening or coverage of waste heaps, it may be cost-effective to carry out a detailed study of soil, weather, etc. This could well involve some field trials[353]. The scene changes as mining progresses[432]. Where the life of a quarry is long, the planting and shaping of areas within the site boundaries to suit the stage of mining should be considered[177]. This could involve planting which will be destroyed in later phases of the working.

The colouring of process plant is the subject of widely disparate views. It is probably more important that the plant should be as unobtrusive as possible and those parts which are visible should be as tidy and well kept as possible. We wonder whether it is wise to agree to a rust/dust coloured finish as this would seem to be accepting that dust and rust will be significant factors.

7. Landscape

7.4.5 Discussion

In the past the approach to landscaping has often been one of providing screens and cosmetic treatment as problems arose, rather than anticipating them at the planning stage. Today there is a more positive attitude but lack of integration can lead to some features being unnecessarily obtrusive and intrusive.

New planting should be integrated with the existing vegetation[396]. Hedgerows can be reinforced by additional planting, and woodlands can be extended. New planting should be of mixed species, merge with the surrounding vegetation and be in sufficient volume to allow for failures. Experiments are taking place in the transplanting of hedgerows, eg by Coal Contractors Ltd and widely by BCOE[456], to provide partial screening and landscaping. The Forestry Commission is contracted by the Department of the Environment to undertake research into the problems associated with the production and establishment of amenity trees. The results of this and other work are channelled through the Department's Arboricultural Advisory and Information Service which is available to help and can supply many notes and other publications[328].

Some operators[384,428] carry out considerable planting prior to submitting planning applications so that the vegetation will be established by the time the application is considered. It is hoped that the strength of objections is thereby reduced. Preplanting does not require planning permission but demonstrates good intentions. At least one MPA[400] states that tree screens and shrubs should be planted prior to an application. We accept that not all operators have the financial resources or a land bank to do this, but we think the approach should be encouraged[460]. This would avoid the need delay operations until planting had occurred.

The alternative of granting planning permission ahead of the 'need' for the mineral, giving time for the preplanting to take place and mature prior to working, has not found favour with MPAs.

7.4.6 Future Good Practice

As mine planning, physical and ecological techniques improve, progressive restoration can be increasingly applied to all types of working, including hard rock quarries, and the visual quality improved.

As the framework within which preplanting can be encouraged is developed and the techniques and economics of transplanting are improved, so preplanting will be able to play a greater role in minimising the adverse impact of workings.

7.5 RECOMMENDATIONS FOR FURTHER ADVICE AND RESEARCH

(1) The use of bunds to protect individual or small groups of houses and footpaths is not always what residents and others seem to want. It may be worth further exploration of the pros and cons of bunds and possibly suggesting limitations on heights and distances. It may also be worth identifying typical circumstances in which acoustic fences would be more appropriate than bunds.

(2) Attempts are being made to apply progressive restoration to hard rock quarries. To maximise the visual benefit, progressive restoration needs to be considered from the earliest stages of site planning. A reappraisal and rethinking of quarrying practice and the sequencing of working could identify approaches conducive to this. Inter alia progressive restoration would involve a firm commitment about the depth of the working. Clearly a number of site shapes, depths, strata geometries, etc would need to be examined.

(3) We suggest that the Department look into the issue of pre-planting to see if there are ways in which this approach, including preplanting off-site, could be facilitated and encouraged.

7. Landscape

SUMMARY OF GOOD PRACTICE – LANDSCAPE

Good Practice for Mineral Planning Authorities

Consider the need to agree or specify planning conditions relating to:
- the sequence of working,
- progressive restoration procedures,
- preplanting and planting requirements,
- the siting of plant and its visibility,
- geometrical screening and the nature of landscaping,
- the location and shape of soil and overburden mounds and waste heaps,
- the use of conveyors,
- the treatment of haul-roads.

Good Practice for Operators

- have a positive approach to the landscape,
- plan ahead for:
    - planting,
    - direction of working,
    - progressive restoration,
    - siting of process plant,
- pre-plant, if possible before making a planning application,
- seek to agree landscaping requirements with the MPA and only depart from them by agreement,
- ensure that site-engineers have, and are seen to have, the will to produce a visually acceptable operation,
- have a good-housekeeping policy, keep the site tidy and well maintained, including paintwork.

CHAPTER 8

GROUNDWATER

8.1 GENERAL

Groundwater and surface water will be dealt with separately in spite of some overlap, for instance dewatering affects both. Dewatering and to a large extent increased drainage is dealt with in this Chapter. Surface water is discussed in Chapter 9.

The principal changes in the groundwater regime which may occur due to surface mineral workings are:
- the removal of topsoil, overburden and mineral, and replacement, possibly in combination with imported materials, which may change:
    - the quality of the infiltrating water recharging the aquifer,
    - the timing and relative rates of aquifer recharge and surface water flows,
- dewatering of workings or diversion of surface water courses which may, in taking water from one place and discharging it in another:
    - change the supply of water to abstractions and spring-fed surface water courses,
    - lead to settlement of the ground surface, buildings, etc,
    - change the quality of the water before discharging it,
- discharges from the workings may cause physical and chemical contamination.

Many of the above changes are inherent in a planning permission allowing mineral to be mined, but in many cases the changes can be ameliorated by appropriate operational practices.

8.2 PROBLEMS

Opinions differ on the frequency of groundwater problems due to the effects of surface mineral workings. Problems can be severe but the cause difficult to determine[462] unless monitoring has been thorough and prolonged; weather and other factors can cause changes which are similar.

8. Groundwater

Some see the difficulties as rarely serious[442] compared to the effects of any subsequent use of the site for waste disposal[443]; others find frequent problems[455,461].

The workings, apart from the effects of dewatering which are considered later, may:
- intercept a perched water table[151,446] and cut off base-flow to springs, streams, etc,
- remove part of an aquifer,
- remove a natural filter medium[447,448] and increase the risk of contamination of an underlying aquifer,
- decrease aquifer recharge and increase surface water run-off[455] where topsoil is removed leaving bare rock surfaces; conversely the removal of vegetation at sand and gravel workings can increase the recharge rate[443],
- obstruct water flows, due to inappropriate backfilling with clay below the water table[443], this reduces drainage 'upstream' of the site and vice versa 'downstream',
- mix two groundwater regimes, eg one in sand and gravel with another in an underlying chalk aquifer[443].

This mixing of two regimes can introduce undesirable silt or clay particles into an aquifer and seriously affect potable water abstractions which do not have treatment facilities. More complex chemical problems arise where there is organic matter within the mineral; this can give rise to acidic or reducing conditions and the mobilisation of metals in either of the aquifers. Working in a gravel aquifer may contaminate it with bacteria from animal life in the working or a surface watercourse; this can lead to bacteriological contamination of an underlying aquifer if there is any mixing of the aquifers.

Examples of problems which result from dewatering[124,165,227,325] and the diversion of rivers and other watercourses include:
- instances of derogation:-
    - the reduction of flow or drying-up of groundwater supply to springs, wells[325,371] and the lowering of flows to and levels in lakes and streams with consequent effects on amenity, dilution of discharges, fisheries, conservation and recreation facilities[443],
    - widespread lowering of groundwater levels leading to derogation of watercourses and failed springs and base flows lower and lasting longer than in the past[455],
    - dewatering of gravel workings has been reported[445,446] as causing derogation of abstractions and loss of crops and fish[445],
    - cutting of perched water tables[446],
- the excessive draining of land with adverse effects on vegetation, agriculture[371] (moderate extra drainage can sometimes be good for agriculture),
- there is some evidence of adverse effects upon growth and diversity of woodlands adjacent to quarries but this may be due to dust and exposure as well as increased drainage[248],
- a reduction in surface run-off from surrounding land, a decrease in evaporation and an increase in the infiltration of rain[185],

- damage to adjacent peat beds/raised mires both in terms of the resource and scientific/wildlife value[15,169]; destruction of previously waterlogged and hence preserved archaeological remains[241],
- ground movement and subsidence[280,443] of houses and roads, especially those adjacent to peat workings,
- slope instability, which in turn can cause damage to neighbouring property, crops, etc and river bank failure[304],
- saline intrusion[445] and drawing in contaminated water[325] from adjacent areas,
- visible precipitation, eg iron, caused by the discharge of groundwater into surface watercourses with differing pH or dissolved chemicals; this may affect wildlife[448].

Pumped water from dewatering is not always used to advantage. Instead of being used to reduce the effects of groundwater lowering[443], pumped water is usually fed to the nearest stream which increases the risk of flooding[321].

The problems of water and physical or chemical contaminants[386], which pass through the site can be:
- further reductions in aquifer replenishment and a flashier run-off due to the increase in impervious surfaces, eg rock, hardstandings,
- the clogging or pollution of aquifers and surface water courses by:
  ○ suspended solids, eg mud, quarry fines, dust removed from processed mineral,
  ○ highly alkaline slurry from wet dust collecting systems on coating plants[455],
  ○ contamination from vehicles running through water on the quarry floor,
  ○ solvents used in processing and spillage of noxious chemicals, especially oil during refuelling[165,446,447,455],
  ○ contamination by waste water discharged directly or indirectly into the aquifer,
  ○ contamination by salts in overburden if it is inappropriately back-filled[104].

Where working starts many years after planning permission has been granted[446], the impetus generated during the consideration of the application may be lost and operations can begin without any agreement on the methods of working to protect aquifers and surface water.

Difficulties in obtaining new consents could increase the trend towards larger and deeper workings. Deeper workings may create more groundwater problems and place limitations on restoration possibilities. Increasing the depth of a quarry may be difficult near to caves which could provide a route for pollution into the groundwater[353].

8. Groundwater

8.3 ACCEPTABLE CHANGES AND MONITORING

The overall requirements are to avoid significantly adverse effects upon the groundwater regime in terms of quantity or quality as set out in the EC Directive on pollution of groundwater[142] and UK circulars[143,144]; MPG7[69] refers to pollution from mineral wastes. The issues that need to be considered are:
- aquifer protection,
- pumping for dewatering, etc,
- recharge or discharge – quantity, quality, rates and sites,
- monitoring,
- potentially affected abstractions,
- ameliorative measures to be employed.

The detail of such requirements can only be considered on a case by case basis. Where conditions are necessary, they should be set by the MPA after consultation with the NRA.

The information available on groundwater is often limited to that obtained during geological investigations within the proposed site. Where adjoining land is not in the operator's control investigation may be difficult and information limited. Even within a site the information is often limited to 'struck' water levels obtained during borehole surveys. The installation of piezometers is simple and would provide more reliable data.

In many cases, monitoring of groundwater in the vicinity of mineral workings is essential so that the operator, the NRA and the MPA are able to:
- understand the pre-existing regime,
- agree the proposed method of working,
- decide upon any conditions to be applied, and
- establish conformance with them,
- provide evidence for future proposals for extension of the workings.

The factors that may need monitoring include:
- groundwater levels around the site,
- state of abstractions,
- quantity and quality of pumped discharges,
- flow of springs, etc,
- quality of groundwater,
- use of preventative and ameliorative measures.

The NRA has regulatory powers to license abstractions. These powers do not extend to the dewatering of mineral workings for operational purposes, however the use of such water for processing may require a licence under the Water Resources Act 1963. The NRA may shortly have powers to control the design and construction of oil storage installations in regulations under the Water Act 1989.

In difficult cases, monitoring may be needed over several years to average out natural variations and establish sufficient data to form a reliable base for prediction and the agreement of working practices. If necessary, planning permission may have to be given with the condition that work does not commence before monitoring is complete and agreement reached on those practices which will be acceptable. However there may be problems of access to suitable monitoring points, especially before planning permission is given.

8.4 GOOD PRACTICE

8.4.1 Monitoring and Prediction

As implied earlier, the greater part of the impact is determined at the planning stage. There is pressure from NRA[443] to maximise the use of minerals from non-aquifer sites. However, there is a need[461] to establish base line information on both groundwater and surface water regimes prior to any working. Data on water levels, flows, their fluctuations and water quality should be gathered both from inside and outside the site. Without such information the understanding of the water regime will be limited and the effects of the workings difficult to determine. There will also be no basis from which to judge future changes possibly related to the workings.

There is a need to consider long-term monitoring of the pre-operation situation including dewatering tests over a period long enough to determine cause and effect. Such studies may be increasingly required before planning applications can be determined[462].

Associated with a recent planning permission for the deepening of a quarry, there was a legally binding agreement to monitor 15 local water sources and to make good any derogation caused by the quarry[455]. This was considered necessary because the Water Act 1963 does not protect a right to underground water but water from a 'defined channel'. Common Law action is the only option[428] available to a plaintiff.

We consider that the installation of piezometers would in many cases provide reliable and valuable information about the groundwater situation both on and off the site. This would permit the situation to be monitored as a working became deeper[124] and problems to be anticipated. Water flows could be controlled and if necessary deeper working stopped to avoid or limit adverse effects.

The techniques of prediction with regard to the design of dewatering systems seem to be well advanced. There is a good understanding of the possible effects[227,280] and of ways to avoid

8. Groundwater

problems. However, the level of confidence in the predictions seems low[6,277] and this may be because often there does not appear to be:
- a good understanding of the base-line situation and data on which predictions can be based,
- a significant amount of operational monitoring to check predictions.

Pumping tests have been suggested[280] where it is thought that there may be effects on the yield of wells or pollution problems.

Particular concern has been expressed[462] that, where proposed mineral workings are near an SSSI or Local Nature Reserve, the consequences of dewatering could be very serious and that a full hydrological study should be required before approval is given. Such a study could use a fluorescent dye to trace water pathways[353] but would need to be carefully controlled in consultation with the NRA.

Against this background, mitigation proposals should be examined prior to planning permission. If appropriate, some measures could be embodied in legally binding agreements. Monitoring during operations would help to anticipate problems, identify the causal factors and indicate requirements for further mitigatory or compensatory measures.

8.4.2 Control Measures

A successful balance between water resource requirements and operational needs is difficult to achieve[407]. The potential for conflict is considerable but difficulties can be overcome by consultation, good planning and design of the operations[456].

Suggestions for amelioration are to:
- avoid dewatering by working the minerals wet[6,445,448],
- leave an undisturbed layer of mineral over an underlying aquifer to avoid contamination, eg with sand and gravel suggestions are 0.5 m[460], 1 m[443] and 3.0 m[384] to provide adequate filtering. The permeability of strata vary considerably and any limit must be determined on the basis of the local conditions. Undulations of the interface between the strata mean that safety margins will have to be added in sensitive areas,
- limit the size and depth of a quarry[165] to minimise problems,
- leave adequate margins between the workings where dewatering is essential and nearby buildings, surface water courses, farms, areas of ecological interest, eg:
    - for peat workings this is especially important[447], a 50 m margin from main drains and 40 m from roads have been suggested,
    - for sand and gravel a 4 m margin from hedges, 10 m from trees and ponds[383], and 10–30 m from watercourses depending on the situation[443],

- provide compensatory measures where there is doubt, eg one study[384] suggested that at 200 m an abstraction was 'safe' from contamination but a filtration plant was offered to be doubly sure.

Sand and gravel can often be worked without dewatering although operators may be reluctant to do so because wet working may recover less of the mineral; dry working is not necessarily harmful[367].

At a more detailed level the planning and design of operations should consider:
- consultation with interested parties; it has been said[442] that potential problems of pumping are well understood and with good consultation can be overcome to everyone's satisfaction,
- the codes of practice[329,330] published by the Federation of Civil Engineering Contractors for temporary spoil mounds and the stability of slopes on opencast coal sites associated with dewatering and drainage,
- dewatering shallow workings progressively in cells[367,371],
- limiting water ingress by using a barrier/seal, eg a trench backfilled with impermeable material such as clay[371], or 'expensive' injection methods for sealing porous strata around the perimeter[185,259,277]; provision for water diversion around the site may be needed,
- providing for archaeological remains to be wholly or partly surrounded by a bund keyed into underlying impermeable strata, and kept waterlogged or screened from the effects of dewatering,
- protecting water courses by trenches backfilled with impermeable material[20],
- storing water to replenish ponds[383] and feed streams in summer which normally rely on springs fed by groundwater[124,447],
- siting oil and chemical storage tanks on impervious bases surrounded by bunds to contain spillage and prevent contamination of underlying aquifers[443]; the bunding should have a capacity of 110% of the storage. All draw and fill pipes should be enclosed[361], tamper-proof taps and valves used[383] and any spillage collected and disposed of carefully[361],
- treating wet-processing plant similarly to oil and chemical storage[448],
- reducing the possibility of leachate from disturbed and back-filled overburden containing undesirable salts, etc by:
    ◦ diluting the problem overburden with other innocuous overburden,
    ◦ protecting it with a clay cap[104] or placing it on an existing clay stratum, eg at the bottom of a sand and gravel pit, and covering it[259],
- controlling run-off, especially for coal, as part of the overall mine water system[227],
- avoiding the return of impermeable overburden to the pit where it may upset the ground water flow[367],
- reducing the effect of impermeable backfill by providing groundwater relief drainage around the site.

8. Groundwater

8.4.3 Compensatory Measures

In view of the often suspected but unproven adverse effects of dewatering, good practice should include precautionary measures, eg the provision of alternatives supplies for abstractions.

Where groundwater supplies in a region will be reduced by a mineral working, the possibility of compensation for this reduction should be explored[124]. An operator surrendering rights to mine below the water table in another part of a region is one way of doing this[200].

8.4.4 Future Good Practice

It has been suggested that consideration be given to:
- recharging aquifers[6,280,445,475] in the area surrounding the mine by pumping into soakaways, injection wells, lakes, or recharge area separated from the site by mounds; there should be monitoring boreholes around the perimeter of the site and the quality of any water for recharge must be strictly controlled,
- recharging via surface water courses; in the Mendips, in order to avoid the danger of reverse flow, it has been suggested[371] that water be pumped 2-3 km to find sink points; complementary to this may be the need to seal the beds of streams which flow over fissures connected to a quarry.

8.5 RECOMMENDATIONS FOR FURTHER ADVICE AND RESEARCH

(1) NRA should be encouraged to publish explanatory notes about their requirements and guidelines on acceptable practice (some limited advice will be available in the National Groundwater Protection Policy when published).

(2) There is little published definitive evidence on the effects of mineral working on groundwater[371] and it seems that more monitoring is needed to establish the extent of the more problematic effects. Operators should be encouraged to monitor the water regime themselves both off- and on-site.

(3) There is a dearth of documented knowledge about the effects of dewatering on flora and fauna, especially in SSSIs, and research is needed[460,466].

(4) The pressure for deeper quarries is likely to increase, so research into the likely effects upon groundwater is needed to provide a basis for the consideration of development proposals.

(5) The dry working of sand and gravel is often preferred by operators when it is possible, even though it may cause off-site problems. Consideration should be given to a study of the comparative cost and benefits of wet and dry working (see also Section 7.4.3). It has been suggested that this might involve a number of site-specific case studies[460].

8. Groundwater

SUMMARY OF GOOD PRACTICE – GROUNDWATER

Good Practice for Mineral Planning Authorities

- have regard to NRA's policies on groundwater protection at the inception and formulation or modification of Minerals Plans,
- after close consultation with the NRA, consider the need to agree or specify planning conditions, to support the protection of aquifers, relating to:
    - delaying operations until monitoring data are available to demonstrate the absence of problems or allow precautionary measures to be agreed,
    - nature, area and depth of working,
    - arrangements for recharge,
    - means to minimise problems from storage of oil/chemicals,
    - monitoring of quantity and quality of pumped flows from site,
- consider the need to seek legally binding agreements regarding:
    - monitoring off-site, eg of groundwater levels and abstractions,
    - compensatory measures, eg for abstractions likely to be adversely affected.

Good Practice for Operators

- consult the NRA at an early stage,
- monitor base-line before design and planning application,
- carry out pump tests,
- monitor during operations:
    - ground water levels,
    - neighbouring abstractions,
    - quantity and quality of recharge flows,
    - neighbouring land, crops, ecology for incipient problems,
- plan to minimise potential problems as well as to meet NRA or MPA conditions,
- consider not dewatering or, if unavoidable, dewatering progressively in cells and reducing the inflow of water by sealing,
- leave effective filter layers between aquifers,
- use codes of practice for temporary spoil mounds and slope stability,
- provide for recharge of aquifers,
- bund waterlogged archaeological sites and provide water supply,
- provide impervious bases and bunding for oil/chemical stores and wet-process plant,
- avoiding seepage of contaminated run-off through floor of quarry,
- encase polluting backfill in impermeable material or dilute it with innocuous fill.

CHAPTER 9

SURFACE WATER

9.1 GENERAL

The main potential effects of mineral working on the surface water regime are to:
- alter the surface over which water flows,
- change the pattern of surface water flows,
- change the quantity and physical and chemical quality of those flows.

9.2 PROBLEMS

9.2.1 Incidence of Problems

The number and seriousness of problems vary significantly with the depth of working compared to the water table, the extent of dewatering, the nature of the mineral and overburden and the method of working. The main problems are changes in surface water flows and their contamination by particulate matter.

In addition, operations near to watercourses can affect access for maintenance[446] and may endanger the stability of their banks[447,474].

As yet there is not enough monitoring to be fully aware of the frequency, extent and magnitude of problems. Certainly, serious problems do occur but it seems to us that there is a tendency to adopt a 'crisis management' approach rather than a planned one.

9.2.2 Changes in Surface Water Flows

Greater and quicker ('flashier') run-off[455], especially in storms, from de-vegetated or impervious areas (rock surfaces, plant hardstandings) and uncontrolled pumped discharge of surface water and groundwater may upset land drainage[448], overload and erode receiving watercourses[447]. These flashier flows lead to increased flows in winter and after storms and reduced flows in summer. Many sumps are not big enough to compensate and there are regular difficulties in maintaining flows in nearby surface water courses during the summer and in sustaining the cleanliness of water discharged from dry-worked pits[443].

9. Surface Water

Surface water courses that are cut by workings and not diverted have to be controlled artificially. If the balancing reservoirs and the pumping arrangements do not effectively mimic the original natural conditions, the downstream regime may be adversely affected by changes in the timing and level of fluctuations in the flow. In one case, a stream was subject to scouring because water was only discharged at night when the pumps could use cheaper electricity.

Overloading of surface water courses with large volumes of pumped groundwater and/or storm water run-off may cause flooding downstream[165]. Ensuing erosion will further augment the level of suspended solids. Together with the other changes to drainage, storage and water courses, silting may increase the potential for downstream flooding[446].

Waste materials deposited and sub-soil or top-soil temporarily stored in washlands can also adversely affect the flood regime[443].

9.2.3 Contamination by Suspended Solids

The discharge of water with levels of suspended solids which are damaging to the receiving water course is widespread.

Silt lagoons are not always effective[443] in removing the suspended solids from surface water run-off, process water or lorry washing effluent. They are sometimes short-circuited as the result of bad design or adverse weather[448]; strong winds can agitate lagoons or storm flows pick up previously settled material. The suspended solids in run-off from limestone and basalt settle out easily but that from sandstone[447] and china clay[316] can be troublesome.

There is major risk of site run-off water containing large quantities of suspended solids[183,448] especially during soil stripping and soil replacement before it is vegetated[14]. Storm flows are difficult to settle and many workings discharge contaminated water in wet weather[455]. It is argued that the receiving watercourse will be heavy in silt at this time anyway, but this is not always the case, eg with spring-fed streams.

Lagoons are not always effective in removing clay, coal and limestone fines[447,448]. Run-off from clay workings is especially contaminated with fine clay, which is difficult to settle and pollutes the receiving watercourses[447]. Drawing process water directly from settling lagoons can 'unsettle' it and create turbidity, ie a loss of clarity, if not a significant level of suspended solids. Lagoons with significant depths of sediment are potentially dangerous, especially to children.

Suspended solids can transform freshwater fisheries[391]. They can directly reduce the survival rates, growth, resistance to disease, feeding and migration behaviour. Indirectly, fisheries are affected by reductions in the food supply. Turbidity damages aquatic life by reducing the amount of sunlight and causing some avoidance action by migrating fish[220]. Heavy turbidity can prevent migration of salmonids[447]. Suspended solids can have a 'choking' effect on fish and invertebrates, and gills and tissues may be abraded[447].

9. Surface Water

The blanketing of a river bed by inert solids can[220,445,446,447]:
- harm fluvial habitats,
- damage fish, eg salmonids, and spawning grounds by making them unsuitable or smothering eggs already laid; over a period this would alter the age structure of the fish population and put fishing at risk,
- smother vegetation,
- smother the habitats of smaller organisms on which fish feed and reduce the biological quality of the watercourse,
- change the nature of the bed from stony/gravelly to silty by repeated deposition and introduce silt-loving species of plants,
- as referred to above, restrict flows and increase the risk of flooding[447].

Suspended solids in the form of fine clay can, even at levels less than 15 mg/l, reduce visibility and sunlight available to flora and fauna in a surface water course. Turbidity can give rise to public concern, especially when highly coloured[447], and cause problems to cattle watering, fish farming and ornamental ponds fed off the watercourse[447]. Discolouration is primarily a visual problem.

9.2.4 Chemical Contamination

A major problem was reported[444] following a rise in groundwater level after a dry period which leached metals into a stone quarry. The contaminated water was discharged into a surface watercourse and killed the fish downstream.

Lagoons used for chemical treatment as well as settlement are said to suffer from a lack of a scientific approach to design and to be poorly managed[447].

Erosion apart, most contamination is from processing. A number of problems have been met in which oil and tar from roadstone coating plants have contaminated surface water. Potential pollutants, eg:
- salts, especially soluble chlorides and sulphates[446],
- lignite in sand[447],
- red marl in limestone[447],
- ammoniacal nitrogen in peat[447],
- minerals, overburden, oil fuel, detergents, reagents, dust[448]

may lead to changes in:
- pH,
- colouring (not usually toxic),
- other problems and possibly toxic effects

for any watercourse into which the effluent from the site is discharged. Flocculants are said to be rarely a problem[220].

Environmental Effects of Surface Mineral Workings

9. Surface Water

Pollution of streams in one region is said[455] to be caused mainly by water contaminated with mud and quarry fines, oil, solvents, etc, from plant areas. It is often caused by vehicles working or running through water on the quarry floor. Poor 'housekeeping' can also be a factor, eg, not clearing up spillages of dust or fines.

The storage, handling, transfer of oil and chemicals and the refuelling of vehicles gives rise to the risk of spillage and is of concern[447,448]. Even small amounts of oil in the discharge can be odorous and adversely affect cattle watering, large amounts can kill vegetation and fish[447]; spillage of creosote-based toxic substances has similar effects.

Groundwater discharged to surface water courses, because of the differing chemistries, can cause chemical precipitation, colouring and effects on wildlife/fisheries[448]. Acid or saline run-off can occur from old colliery spoil. Dewatering discharge from gravel workings directly colours and blankets the beds of receiving streams[445]. Acid drainage[105,177,186] from pyrite in coal (only likely from the reworking of old deep-mine tips) or from sandpits[447] can cause corrosion, discolouration, decrease oxygen levels, limit flora and fauna and prevent recreational use.

9.3 ACCEPTABLE CHANGES AND MONITORING

In practice nearly every discharge from a mineral working is subject to statutory control under the Water Act 1989 and needs prior consent from the NRA. Planning conditions may reinforce these requirements but the primary control is with NRA.

Limits on discharges from dewatering and processing are set on a case by case basis. No overall guidance is available. The use of flocculants and other chemicals will need to be discussed with NRA in advance. Methods of measuring turbidity are available but are rarely used or limits set.

Ideally no deterioration in the quantity or quality of the surface water would be permitted. In practice baseline monitoring may not have taken place or, if it has, not over a long enough period to ascertain the existing levels and variations in the flow and quality. It sometimes appears difficult to frame and adequately enforce limits related to storm conditions. Some operators believe that they are 'entitled' to relief from absolute limits in such circumstances without the prior agreement of NRA.

Depending upon the existing levels of contamination, the capacity of watercourses to accept changes of flow and additional contamination varies enormously. It may be argued that a polluted water course can accept a more polluted discharge on the basis that it cannot do further harm. Depending upon the time-scale however, further pollution may delay clean-up efforts.

The principles of legislation and limits are much discussed[132,187] but the literature relating to acceptable levels of surface water quality in the context of surface mineral workings is sparse.

9. Surface Water

It is reported that china clay in suspension alone is acceptable at 25 mg/l whereas more than 400 mg/l would cause fisheries to be 'poor'[391].

Diversions or alterations to the course or bed of a river may require the prior consent of the NRA under Land Drainage Act 1976 byelaws. The NRA also has a responsibility to protect and enhance the conservation and amenity of river corridors under Section 8 of the Water Act 1989.

9.4 GOOD PRACTICE

There is a large volume of literature on control[6,46,54,72,112,118,159,165,166,177,183,250]. It is said[447] that technical solutions are available for most potential pollution problems.

Surface water demands much attention when designing mineral operations. The principles underlying the potential problems of pumping and run-off are said to be well understood and given good consultation can be overcome to everyone's satisfaction[442].

Most potential problems can be dealt with by good design of the water system. This should integrate the various requirements and optimise the cost-effectiveness of the system. It should be implemented with the appropriate monitoring to confirm its satisfactory performance and to anticipate problems[177,183,316]. Contamination from the excavation itself can be limited by restricting the water flow into it[118], diverting water courses and bunding of the excavation. Water pumped from the pit should be kept separate from surface water and treated if there is any risk of contamination[165].

The baseline situation regarding levels, flows, fluctuations and quality in surface water courses both inside and outside the site[461] should be established before working starts. There is a general need for baseline data[406]. It would help if everyone concerned with surface water were to keep records, eg of periods when steams are dry, of levels in wells. As stated in Chapter 8, there is also a need for the monitoring of dewatering tests.

9.4.1 Control of Water Flow

Surface water coming into the site should usually be diverted away from the working areas[72] or culverted so that contamination is avoided.

Flow balancing of run-off and dewatering discharge should be used to avoid upsetting local surface watercourses and drainage[443,448,455]. Flumes can be set up to measure flow and to activate pumping when appropriate. Pumping from excavations into watercourses must cease when they are running full[443].

Settlement lagoons can be adequate to provide flow buffering capacity in the event of storms[445] but usually they are not. Where the site is in a low amenity area and the receiving

9. Surface Water

stream has sufficient dilution capacity[448], a bypass can be provided[166] to avoid scouring of the lagoon during flood conditions. A better approach is to have a very large sump in the quarry floor with floating pumps[455] to ensure that water extraction is as far as possible above any sediment[183]. It is not always easy to have a large sump, eg when deepening a quarry.

It has been pointed out[455], however, that flow balancing and recharge reservoirs, pumping stations and pipelines have adverse effects, eg visual and land-take, which need to be considered as well as beneficial ones.

The advantages, in addition to flow balancing, of storing water on site are[407]:
- storage of rainfall,
- provision of a sediment trap,
- easier water treatment,
- provision of water for dust control and processing; this may need an abstraction licence under the Water Resources Act 1963.

Sand and gravel[6,46] processing water is often drawn from the pit and then mostly recycled through settling ponds with little, if any, discharge off-site, ie, effectively a closed circuit system. An alternative to using a sump on site for flow balancing is to use water from an off-site source.

9.4.2 Control of Erosion

The first approach in the control of physical contamination is to minimise the pick-up of sediment by erosion.

Control of erosion and run-off[54,112,159] has a number of elements and uses a number of 'tools', eg vegetating surfaces, mulching, straw bales, roughening/ridging of surfaces, terracing and low gradients, netting and other geo-textiles. Vegetating surfaces can also aid slope stabilisation and protection[480]. In varying degrees these measures lessen the disturbance due to rain, reduce erosion by protecting the surfaces and improve the settling-out of sediment by slowing flows.

The flow from 'overload' run-off should be canalised in a stable way, eg by a perimeter or cut-off ditch[166] or on-site drainage[34,50]; a settling/stilling pond system will probably be needed[72]. A cut-off ditch gradient of about 1:40[166] is sufficient to ensure drainage and avoid ponding, but low enough to avoid scouring.

Swift restoration, especially when it is done progressively, minimises the problem of erosion.

9. Surface Water

9.4.3 Removal of Suspended Solids

Solids in suspension will need to be settled out before discharge. This is usually done by sedimentation[183] in separate lagoons but much can done in a large sump. The lagoons must be of sufficient size[447] and, if long and narrow, sited at right angles to the prevailing wind[447] to avoid undue agitation and mixing.

A lagoon, or better a pair to allow them to be emptied of sludge in turn, is usually adequate to control sand and gravel washings[445,447] and run-off from limestone and basalt[447]. Lagoons should be sited away from abstractions[367]. Where dewatering is performed on an area by area basis, excavated compartments can be used as lagoons for subsequent working compartments[367].

The discharge from lagoons may need to have secondary treatment, eg, by a reed bed, to remove residual solids especially from washing plant[367,443]. Lagoons associated with coal workings may need flocculants or supplementary filtration[183].

The settlement of all quarry effluents is assisted by the use of polymer flocculating agents[447], eg polyethylene oxide(PEO)[420]. Their use should be discussed with the NRA. Clay may become colloidal[288] and need a positively charged additive to cause flocculation, eg an organic polyelectrolyte.

Some operators are considering the use of mechanical means to remove solids[455], eg screens or presses in combination with pretreatment by flocculants[419]. Another possibility for post-flocculation treatment is the use of a direct electric current[418] to separate fines from water.

9.4.4 Control of Chemical Contamination

The control of chemical contamination needs to be considered and generally need be no more than a pH correction in conjunction with settling lagoons[446]. Contamination, eg by chlorides, can be dispersed by dilution in larger volumes of water[446]. Interceptors may be necessary to retain oil-based substances[183] especially on sites with a lot of plant, eg opencast coal. A boom is one possibility which has the advantage of picking up any 'scum'[447].

Sewage should normally be discharged to the sewage system but can be discharged via septic tanks, subject to obtaining NRA consent[183].

Contamination by oil, chemicals and wet processing on site can be limited by ensuring (see Section 8.4) that:
- storage is outside the operational area,
- liquid storage is over a bunded impervious surface,
- spillage is collected and disposed of appropriately.

9. Surface Water

Acid contamination from coal waste tips, which is due to the oxidation of pyrite, can be controlled by a rapid build-up of the tip and by early covering[118]. This is only relevant where old deep-mine tips are disturbed or worked and does not occur with opencast coal overburden mounds[456]. Contaminated run-off must be collected and treated before discharge[183].

For coal[183], associated with iron, pH can be corrected by the addition of calcium hydroxide, 'milk of lime', sodium carbonate or sodium hydroxide solutions which precipitate the iron. In difficult situations, supplementary filtration may be necessary. If heavy metals are present they will need careful scientific consideration and treatment. Alternatives to neutralisation of acid contamination are ion exchange and reverse osmosis[105].

Chemical contamination of processing water can be controlled[54,250] by tailings lagoons, flocculants, chemical balancing of pH, dispersion, dilution, pumping or permanent containment. Much can be done by careful design of the water cycle[183].

9.4.5 Control of Other Potential Problems

Measures to control other problems include:
- leaving a generous margin between neighbouring watercourses and mineral workings[364,443,447] to avoid undercutting, and to provide a working area for maintenance,
- consulting with the NRA about the environment of any nearby river corridor,
- avoiding deposits of overburden, waste, sub-soil, etc on flood plains but, if they are unavoidable, leaving them there for the shortest time possible and aligning the piles or leaving gaps[475] between them to provide the minimum impedance to flood flows[443].

9.4.6 Future Good Practice

Problems which occur near to surface mineral workings would be more readily resolved with better records and understanding of pre-existing water regimes. There is considerable uncertainty about the appropriate way of providing for storm conditions. At the same time, there is potential for a much higher level of sophistication in the design of water systems in terms of settling out solids by mechanical, chemical and electro-chemical methods.

There is great scope for improvements in practice as more experience and understanding is gained of existing water regimes, the requirements for surface water discharges and the means of treatment.

9. Surface Water

9.5 RECOMMENDATIONS FOR FURTHER ADVICE AND RESEARCH

(1) The NRA should be encouraged to publish explanatory notes about discharge consent conditions, practice concerning alterations to surface water courses and the issues relating to the environment of river corridors. In particular, there is uncertainty about how storm or prolonged wet weather flows should be handled. It appears that some operators find it difficult to accept that a limit applies whatever the conditions unless there is a specific relaxation.

(2) The fact that difficulties arise in spite of an apparently comprehensive range of control measures, suggests that guidance is needed on problem avoidance. The NRA is still relatively young and its relationships with MPAs and operators are still developing. Codes of practice may be beneficial.

For example, guidance from NRA on the design and management of settling lagoons would be useful in view of its suggestion that many lagoons are not scientifically designed. It has been suggested[466] that it is necessary for there to be research into the performance of settling lagoons before guidance can be offered, we are not sure. We accept that NRA need not and should not take any responsibility for the design of facilities to meet the requirements that it specifies, however its involvement in the framing of codes of practice would be helpful.

(3) There appears to be a need for more baseline studies. These should include subsequent monitoring of water quality, and the effects of changes in flows and water quality on flora and fauna. Such studies would benefit from being combined with similar ones on groundwater and dewatering. It has been suggested[466] that irrigation by 'seepage' and 'drip' methods using water from workings should be studied.

As the remit of this programme of studies is so broad, it should be preceded by a short study to determine detailed terms of reference, costs, possible collaboration and funding. It should formulate a framework within which site-specific studies could take place, and identify the terms of reference for the individual site-specific studies.

9. Surface Water

SUMMARY OF GOOD PRACTICE – SURFACE WATER

Good Practice for Mineral Planning Authorities

Consider the need, after consulting the NRA as statutorily required, to agree or specify planning conditions relating to the:
- siting and landscaping of flow-balancing reservoirs,
- siting of settlement lagoons and disposal of silt,
- siting of overburden mounds and waste heaps,
- provision of hard standing and bunding of storage/process areas,
- diversion of water courses,
- provision of monitoring.

The requirement for the operator to have consents from NRA for alterations to water courses or for discharges will cover many issues.

Good Practice for Operators

- consult the NRA about any alterations to existing surface water courses, nearby river corridors and any fixed discharges,
- undertake a baseline survey and establish a monitoring system,
- provide bunding to keep surface water out of workings,
- design water system, including dewatering flows, in an integrated way covering:
  - flow balancing by sumps and pumping,
  - quality control by settlement in sumps and lagoons, pH balancing,
  - oil and scum removal,
  - use of water in processing plant and treatment of effluent,
  - containment of spillage from storage and processing areas,
  - use of water in dust control (may need abstraction licence even if resulting from dewatering),
  - use of clean water to counteract groundwater lowering, eg in nearby pools,
  - regular cleaning and maintenance of water system,
- limit erosion by:
  - rapidly vegetating exposed areas,
  - vegetating, physically protecting or roughening the surfaces of overburden, soil or waste mounds,
  - progressively restoring working areas,
  - lining water courses,
- design sumps and lagoons to cope with all conditions, including agreed or specified storm return period, by ensuring that:
  - they are big enough,
  - scouring is avoided,
  - the retention time is adequate, if necessary, enhancing settlement by use of agreed (with NRA) flocculants or mechanical means,
- use progressive working so that previously excavated areas serve as lagoons,
- leave margins around water courses, river corridors and other sensitive areas,
- minimise obstruction of flood regime by mounds of overburden or waste.

CHAPTER 10

WASTES

10.1 GENERAL

Most references deal with the long-term issues and not with the environmental effects of waste during the operational phase of mineral workings. Conversely this report is not concerned with long-term storage or disposal problems.

Waste, for the present purposes, will be considered as any material not removed from the site for a useful purpose; it can be considered as either 'temporary' or 'permanent'. 'Permanent' waste will be regarded as that dumped outside the excavation and which will remain there. 'Temporary' waste is immediately or ultimately backfilled or otherwise utilised within the excavation.

The main effects of waste are to:
- occupy space within or outside the working area,
- be visible,
- be a source of dust,
- be a source of sediment and other contamination in run-off,
- affect the surface water regime, eg by changing surface water flow in a flood plain.

The main sources of permanent waste are china clay workings, some hardrock quarries and slate quarries which produce large volumes of non-toxic waste.

Most sand and gravel workings do not produce much, if any, permanent waste; some produce significant volumes of clay and silt. Silt is in some cases dug from the settling ponds and used during restoration.

Granite and other hard rock quarries[286] may produce a lot of permanent waste especially in winter when clay sticks to stone and makes it unsaleable[455]. The china clay industry has a particularly high ratio of waste to saleable mineral.

Large volumes of temporary waste in the form of overburden are produced by opencast coal mines. At any site, top- and sub-soil can form significant temporary mounds outside the excavations but, whilst they can have some of the effects of wastes, they are irreplaceable materials for restoration and should be carefully conserved and not regarded as waste.

10. Wastes

10.2 PROBLEMS

The china clay industry has particular problems with its very high level of permanent waste which it cannot return to the excavation. The sand tips are unsightly to many people and a few tips have failed[190]. The tips and the disposal of mica tailings represent major issues which have yet to be resolved, although research is currently in hand on behalf of the Department of the Environment.

Otherwise, waste presents a relatively infrequent problem during the operational phase of mineral workings. For the most part, it is a potential problem which can be avoided by good practice. Inappropriately located waste, both temporary and permanent, can give rise to problems.

Waste can be a greater visual intrusion than the excavation[455]. Problems occur mainly when waste is dumped off-site or piled-up above the sky line, especially when it is not landscaped or vegetated. Although it is often used as a visual or noise screen, it can be considered an eyesore.

End-tipping continually creates a freshly exposed, unvegetated and often unsightly surface. This practice is fortunately not as common now that the problem of potentially unstable slopes has been recognised.

Run-off from wastes can cause erosion and carry sediment which contaminates water courses. Suspended solids (and acid drainage) may harm fluvial habitats[27]. Waste may also create problems if dumped on washlands where it may give rise to flooding[26]. Chapter 9 deals more fully with these effects on surface water.

An instance has been recorded of slurry resulting from the use of wet dust collectors on process plant which contained phenols and caused high pH levels when disposed of in a water-filled quarry[455].

Disposal of silt from settling ponds can be a problem. It may not be sufficiently dewatered to support any significant weight and hence be dangerous unless treated. It may also be a source of dust if it dries out. The problems of dust are dealt with in Chapter 6.

Waste which exists on a site before mineral workings start also causes problems. Colliery waste from old deep-mine workings on opencast coal sites can oxidise if exposed to air and, if it is leached into water courses or aquifers, can create acidic conditions. Waste which is fly-tipped, and industrial waste on a previously developed site may be of unknown origin and be potential sources of soil and water contamination.

10.3 ACCEPTABLE SITUATIONS

Ideally all waste should be kept out of sight within the workings and without causing adverse effects, eg contaminating water, creating a hazard due to instability or giving rise to dust.

In practice, it will usually be a matter of agreeing practices to be adopted in handling and the temporary or permanent disposal of waste. Pollution of surface water or groundwater must be avoided[143].

Where waste can be used to backfill the workings without creating problems for ground or surface water or the final restoration, any problems will be temporary. The visual and other potential problems of temporary waste reinforce the benefits of progressive restoration. The acceptability of visual intrusion created by waste heaps will depend on the duration and degree of intrusion (see Chapter 7).

10.4 GOOD PRACTICE

10.4.1 Current Good Practice

Wastes should be used positively wherever appropriate, eg where they would be beneficial as part of progressive restoration, or amenity screens[285].

In cases where the overburden is back-filled and sub- and top-soil reused[42] there may be problems due to the handling and storage of the material. Material tends to become compacted. This may not matter unless drainage is affected within the rooting depth of vegetation, including trees. Top-soil can be reworked but compaction and deterioration of subsoils must be avoided if restoration is to be successful. To minimise long-term damage soil should only be moved when it is dry. Storage should be in areas safe from compaction by traffic, intermixing and erosion. Grassing discourages weed growth and domed heaps reduce water infiltration. Further guidance on the handling of soils is provided in MPG7[69].

Direct placement of waste is likely to be environmentally preferable to storage. Waste heaps should be landscaped and vegetated as soon as possible. We suggest that the ideal is for waste to be immediately incorporated in a progressive restoration or disposal scheme, if necessary, outside the workings.

Various methods have been developed for revegetation, eg hydro-seeding with fertilisers and mulch has enabled the establishment of vegetation on some china clay wastes. Grass swards may benefit from grazing by sheep[40]. The sooner revegetation is done the less likely it is that undesirable products of oxidation will be leached out or picked up in run-off.

Fine waste, eg dust, can be treated by cementation to create non-erodible solid material. Silt can be mixed with other materials or used as part of landscaping and covered with more

10. Wastes

robust materials. It should be confined permanently and securely away from water supply catchment and recharge areas[46].

Run-off should be caught by trenches, then drained to and treated by the site water-treatment system and settling lagoons. Where it is necessary to site waste heaps in a flood plain, the heaps should be arranged parallel to the direction of water flow to minimise the degree of blockage[443]. Where this is not possible, gaps should be left in any long heaps sited across the flow of flood water (see Chapter 9).

In some cases the amount of waste can be reduced. Slate waste can be reduced and production of slate improved by the use of diamond wire cutters rather than blasting[376].

10.4.2 Future Good Practice

In the future, there will be pressure to maximise the use of mineral resources, to reduce the amount of permanent waste to a minimum and to use it beneficially[78]. Planning and economic pressures will surely lead to an increasing value of minerals and to, for example:
- more economic ways of washing and cleaning rock contaminated with clay,
- the greater use of silt, perhaps in horticulture.

Where waste cannot be used or reduced, environmental pressures will lead to it being 'hidden' more quickly, assisted by the development of:
- quicker methods of sealing and vegetation,
- improved techniques of progressive restoration, especially in hard rock quarries.

10.5 RECOMMENDATIONS FOR FURTHER ADVICE AND RESEARCH

(1) The industry itself should develop ideas for the greater use of mined material by the:
- more efficient recovery of useful minerals, eg by better methods of cleaning,
- use of wastes which others might regard as a raw material, eg use of silt in horticulture,
- identification of alternative uses for 'waste'.

The last of these, ie the uses to which waste could be put, is the subject of a current study for the Department[345].

SUMMARY OF GOOD PRACTICE – WASTES

Good Practice for Mineral Planning Authorities

Consider the need to agree or specify planning conditions relating to:
- the location of waste heaps both temporary and permanent,
- means of controlling of leachate and run-off,
- the height and shape of waste heaps,
- surface treatment, eg vegetation,
- progressive restoration, preferably within the workings,
- the period within which temporary heaps must be removed.

Good Practice for Operators

- minimise the production of waste,
- try to find a use for waste, eg landscaping,
- site waste heaps within workings wherever possible,
- use waste as part of a programme of progressive restoration,
- landscape and vegetate waste heaps as soon as possible,
- site waste heaps having regard to potential effects upon:
    - the landscape,
    - groundwater,
    - surface water courses,
    - the flood regime,
- ensure that waste with a physical or chemical contaminant is encased, eg by clay, so that it cannot escape to the atmosphere or be leached to aquifers or surface water courses,
- store top- and sub-soil and overburden, in a manner which is compatible with ultimate restoration.

CHAPTER 11

SEVERANCE AND FOOTPATHS

11.1 GENERAL

Severance can be direct as a result of cutting roads and paths[a] and by diversions making journeys longer and less useful.

Severance can also be the result of more subtle effects. Existing paths may be made less attractive than they were due to dust, noise or by enclosure between fences. As a result they may become less frequented.

The results of severance are to:
- remove access to leisure/informal facilities, eg paths for walking the dog, meadows for enjoyment of flora and fauna,
- make access more difficult, eg by having to make detours to visit neighbours,
- cut-off animals from feeding areas.

Few data have been found.

11.2 PROBLEMS

Footpaths[463], recreational facilities, roads, canals[474] and animal tracks are cut and taken out of use by workings. Diversions may increase distances and make journeys longer[351]. This is a matter to be considered during the planning stage, but there may be requirements to divert or reinstate paths during operations which need to be monitored and enforced.

Better alternative temporary footpaths could be provided[324]. Even where footpaths are neither severed nor replaced they may be surrounded by a 'hostile' environment, eg, edge of quarry, high earth bunds[242], which creates a psychological deterrent and in effect causes partial severance. Mounds are sometimes regarded as 'alien' features in the environment which dominate alternative footpaths[352].

---

a. The terms 'path', etc, are used in a colloquial rather than any legal sense.

Environmental Effects of Surface Mineral Workings

11. Severance

Many of these issues are likely to apply to animals but no specific data has been found.

11.3 ACCEPTABLE CHANGES AND MONITORING

The extent to which footpaths, bridleways and animal runs can be protected and severance minimised is clearly a matter which has to be judged on a site specific basis; it is a material consideration for a proposed working[148].

Footpaths may have to be re-routed if they cannot be maintained along their original alignment. Indeed, under the Town and Country Planning Act 1990, an existing right of way cannot be obstructed until an order for diversion or extinguishment has been made and confirmed[148]. We understand that, in practice, extinguishment is very rare in connection with surface mineral workings. The provision for opencast coal workings is under separate legislation and the use of a path can be 'suspended' during the period of operations.

The issue is therefore primarily the quality of the diversion. The longer the period of diversion and the greater the use of the original path, the higher should be the quality of the diversion. It helps to have baseline data but survey information has to be used with caution. One survey of footpaths[476] found that some were overgrown and over many observation periods no one was seen using them. Whether no one used them because they were overgrown or vice versa was not recorded!

Where access is significantly disrupted, it may be necessary to consider what compensatory measures should be taken to offset the disruption. Whatever is decided, there is a need to ensure that the installation of the alternative facilities is prompt and that they are regularly inspected and well maintained.

11.4 GOOD PRACTICE

As well as a survey of rights of way, a survey of the pattern of usage of paths and tracks by both people and animals may be useful. The survey would help to determine whether or not a path will be missed during the life of the workings and whether the diversion will be an adequate substitute. In significant cases additional facilities may be need to be provided.

Wherever possible paths, ancient highways, or wildlife runs should be rerouted rather than destroyed[384]. They should be diverted before existing routes are destroyed and replaced as soon as practical. However, in the case of hard rock quarries replacement is often impossible. Operators might consider providing new or improved rights of way, circular routes, etc, where demand is likely to exist. BCOE have provided a minibus service[351] where a path could not be diverted.

Where paths cross the lines of conveyors, bridges should be provided[6]. The landscaping/planting of paths along or between bunds or mounds should be considered to make them more attractive.

In all cases footpaths should have[357] clear signposts, good stiles and fences all of which, including the footpaths themselves, should be periodically inspected and maintained with a usable surface at all times[400].

11.5 RECOMMENDATIONS FOR FURTHER ADVICE AND RESEARCH

(1) We understand that codes of practice exist for farmers in relation to footpaths. The Department might consider the production of a code of practice relating to footpaths and animal runs in the context of surface mineral workings.

11. Severance

SUMMARY OF GOOD PRACTICE – SEVERANCE

Good Practice for Mineral Planning Authorities

Consider the need to encourage or agree or specify planning conditions relating to:
- those routes which are to be retained,
- those routes to be diverted, pointing out the necessary statutory procedures,
- the treatment of these routes,
    - physical, eg surface, stiles,
    - landscaping,
    - signposting,
    - maintenance,
- the provision of compensatory measures where no alternative route is available.

Good Practice for Operators

- understand the prior use of their sites by people and wildlife,
- retain existing facilities as far as possible,
- otherwise make alternative provision where necessary and practicable,
- maintain the availability and quality of these alternatives, in terms of:
    - landscaping,
    - surface,
    - stiles,
    - signposting,
    - fencing,
    - safety,
- provide compensatory facilities where necessary.

CHAPTER 12

EFFECTS ON RESIDENTS

12.1 GENERAL

Residents are potentially affected by all the effects so far discussed. Individually the effects can be ameliorated as previously described. However, we believe that it will be worthwhile discussing some of the issues raised by the situation as a whole; the literature taking a holistic view is sparse.

In this Chapter, we express a number of opinions based upon inferences drawn from the literature and, more importantly, from discussions with operators and local authorities. This Chapter is therefore more subjective than others.

There seems to be a feeling on the part of some operators that, provided they do not create an actionable or a statutory nuisance, they are doing as much as can be expected of them. However, the quality of the residential environment would soon be much poorer if each new development were permitted to cause deterioration just short of nuisance. Desirable environmental quality demands much higher standards and MPAs will need to specify and enforce them. Only temporary features of surface mineral workings can be subject to some relaxation in these desirable standards.

Even a low level of impact may be a serious disturbance to the more sensitive sections of the population, eg frail old people living alone and those who are ill. Many 'locals' will see an operator as an 'intruder' and the workings as 'alien' to their neighbourhood.

A Groundwork Trust report[51] stated that the industry is seen as having a detrimental social impact due to insensitive methods of working, a lack of measures to ameliorate effects and a general failure to relate and liaise with the community. Residents suffer without perceiving a benefit. Views are expressed[170,487] that the industry must make even more effort to address concerns expressed by the public. Why should this be in spite of the many things that the industry does to reduce its impact?

To some extent it is artificial to consider effects, eg of noise and dust, in isolation, because they are rarely experienced by themselves. There is a need to consider the cumulative effect of surface mineral operations. This may be more than an 'arithmetic addition' of the effects that have so far been discussed.

12. Residents

We suggest that public reaction may be due, not only to:
- the loss of a familiar landscape, losing both the visual amenity derived from it[463] and the use of it, albeit the existing use may be an unofficial one[472],
- the introduction of environmental disamenities,

but also to:
- a dislike of change and a fear of declining property values or being unable to sell homes.

It has been suggested[110] that the overall effect is to change the spirit of the community and its image. It often seems that there is an underlying and unexpressed desire to keep things as they are; there is a fear of change, particularly when it is caused by an activity which many may see as alien to their locality. The fear of change, which is difficult to verbalise in a way that appears rational, and of monetary loss is sometimes translated into vociferous objections about scars on the landscape, etc, which seem unreasonable to the operator and possibly to others detached from the locality.

There is often a lack of communication and understanding between the operator and the community which arises for a variety of reasons[365]; simultaneously they may be saying to themselves:

OPERATOR	COMMUNITY
Why don't they understand?	Why don't they give us the facts?
Environmentalists are all alike. They are fanatics. Who wants people problems?	We want to be reasonable, why will they not discuss things with us?
A problem ignored is a problem solved.	We will have to take them to court to get any action.
We will solve the problem technically. Why cannot they be logical?	Can we be sure that they will do it?
We have done what we are legally obliged to do. Surely they do not expect more.	It is not right that they come here and destroy our environment.

There are, of course, points on both 'sides'. In some cases, operators may have a greater concern for the environment than some of the more vociferous partisan objectors[343] and may have to communicate their policies and practices more positively[286]. Even an operator with good intentions, standards and supervision can be stubborn and secretive about his experience on other sites and may be unwilling to accept the need for local variations to his normal 'centralised' policy. By giving standard responses, the operator can give the impression of not listening to comment or criticism.

The NIMBY syndrome can also cause objectors to exaggerate. There is also the possibility of there being a political element to the reaction, particularly if the development is seen as unnecessary or benefiting others who are not affected by it.

Where the developer/operator is well known for a poor past performance on other sites there can be a major problem in gaining public acceptance[32]. Conversely, an operator with a good record[401] can find less difficulty in convincing people of his good intentions and that obtaining planning permissions is sometimes easier. Operators need to be jealous of their reputation if they are to obtain and retain public confidence.

It is worthwhile for an operator to keep a systematic record[292] of complaints, of the actions taken and of any follow-up. This helps to ensure that complaints are not forgotten, that complainants are kept in touch with the progress of any remedial works and that there is a record of site performance which can be monitored.

The attitude of residents can be changed both to their own benefit, because they may be less annoyed, and to the benefit of the mineral operator, because he gets fewer complaints and less hassle. The attitude of visitors, commercial and industrial organisations and potential investors is likely to be similar. Possible ways in which an operator, who is prepared to change his attitude, can bring about this change in the community's attitude include:
- forging community links,
- providing information, especially in advance of problems,
- improving his image,
- providing compensatory measures.

Factors which affect the public's attitude to the operator and the level of objections are the:
- operator's attitude towards them,
- information available to the public and their understanding of the operations,
- compensatory measures,
- duration of nuisance, incremental effects of successive operations,
- hours of work.

None of the approaches, which will be discussed below in more detail, will be fully effective unless:
- there is desire to be a good neighbour,
- there is a will on the part of the operator's management to ameliorate the environmental effects,
- the management and staff at all levels are sufficiently trained to be aware of the environmental issues and to be able to deal with them, and
- the community believes in the existence of this desire, will, awareness and ability.

12. Residents

12.2 COOPERATION AND COMMUNITY LINKS

Many operators have found[eg.428] that cooperation, rather than confrontation, with local authorities has proved to be more productive. A view supported by many MPAs[465] is that problems can be overcome by good liaison. The appointment of a 'planning liaison officer' has proved to be useful[467]. He is the initial contact and channel for all communications with the MPA and with residents. He can ensure that any points or complaints are passed on to the relevant technical staff and dealt with.

Good links with the community at large are perhaps less common but are valued by some operators[466]. One community[451] in Northern Ireland exemplifies the frustration that communities sometimes feel; it met with such a total lack of communication that it no longer thinks it worthwhile writing to the operator. In this case, the planning authority did communicate but was perceived as ineffective.

Effective liaison committees[14,27,267,489] have been found to be a great help by MPAs and operators, eg BCOE over many years, in avoiding and remedying difficulties and they are widely supported[400,435]. People do not necessarily believe experts; liaison committees can assist in resolving this difficulty by improving communication and can help:

- to achieve good relations between the operator and the community,
- the operator to be a good environmental neighbour,
- to reduce complaints by increasing understanding,
- to agree actions,
- to increase the acceptability of the workings.

Liaison committees should be run by the locals, but can be provided with facilities and supported by the operator. MPAs should encourage involvement of residents, amenity and rights of way groups. It is desirable that officers and members of all relevant local authorities[400] play some part in such committees, but it is recognised that there may be difficulties with staff resources. District Councils seem to have more opportunity for involvement than County Councils. Experience of the value of the media attending such meetings is mixed. It is suggested both that the value of liaison committees is improved[267] by their attendance and that it inhibits, or would inhibit[455], free and frank discussion[456].

One District Council has suggested that liaison committees ought to be mandatory[427]. In practice, it is not possible to bind public involvement by a planning condition or a Section 106 agreement[362]. Any requirement on the operator may have to be limited to providing the funding and/or facilities[400] for a committee and to encourage others to join.

Liaison committees do not replace effective day to day communication. Operators should provide names and telephone numbers of contacts who will receive complaints and queries[455]. One residents' association[358] said that what it wanted from the operator was a guarantee of remedies to problems without argument or excuse. Whilst the perceived problems do not always result from mineral workings, it seems that too often some operators rely on the response 'you cannot prove that we caused the problem'. In the longer term however, some flexibility may be worthwhile to engender good relationships with their neighbours.

Many complaints, eg of blasting vibration, are influenced by apprehension about safety and a fear of the unknown. They will be reduced by greater understanding and familiarity with working methods. Mineral workings can be interesting and this aspect should be enhanced by providing access for visitors including casual ones, school children and tourists. The provision of teaching aids to encourage exploration of the different problems of operators and the local community should help improve understanding[200]. Visits by neighbours, local councillors, etc, to similar sites at the planning stage[31] and to the site at the operational stage can help considerably. The occasional or permanent 'open house'[203] can also be valuable. One operator invited his neighbours to an open-day and found the response so great that he had to organise half a dozen.

12.3 OPERATOR'S IMAGE

It is clear that the community needs minerals and the industry needs the community to be reasonable about its difficulties[32,76], so there is a mutual interdependence. It is a considerable help to the industry if the community believe it to be 'socially responsible'. Some operators however seem unconcerned and wait to be taken to task by the regulatory authorities[429].

A combination of good site planning and management should be concerned with for all aspects of surface mineral working. However, the informal comments that we received suggested that some operators and their site managers were inward looking and did not concern themselves with the public's perception of them.

The public's view of an operator in one locality may be reflected in the success or otherwise of a planning proposal in another area[338]. The image of the minerals industry as a whole is also relevant. Individually and collectively the industry will find a good record benefits future planning applications.

The 'About Face' programme in the USA[37,44] for quarry operators appears to be valuable as a means of winning more 'sympathy' from local communities. The programme[207] centres on the visual appearance of quarries but also includes:
- erosion,
- stock and waste control,
- dust,
- settling ponds and sediment,
- loading facilities, and
- lorries.

There is little doubt that good public relations, as opposed to 'pulling the wool' over people's eyes, reduce complaints. A pride in the appearance of a site, in keeping equipment clean

12. Residents

and well maintained, is claimed to help working morale, efficiency and safety[30,31]. In turn this means that workers convey a positive image to their neighbours and visitors[203].

In the UK, the 'Brightsite Campaign'[302] is similar to the 'About Face' programme but is primarily concerned with the visual appearance of all industrial sites. Noise and vibration, dust, windblown refuse, noxious fumes, traffic volume and relationships with the local community are also factors. It is different in that whilst sponsored by industry it is run from outside by the Groundwork Foundation.

A small operator involved with the community and employing local workers may be subject to less pressure than a larger more remote one. The local community are likely to be better informed about the activities and more concerned about the socio-economic consequences of any objection or complaint that they might make. There are, of course, large operators who also recruit the major proportion of their workforce locally and who successfully integrate with the community.

For longer term improvement of image, operators should consider upgrading their existing process plant, moving it to a less sensitive location and modernising it[374]. Ultimate restoration also has its effect on the image of operators, in a less direct and immediate way. Schemes promoting this include the Sand and Gravel Association's awards which started in 1970[341,460]. National sponsorship of such things as nature conservation[412] also helps to improve the image of the industry as a whole and of the individual operator. The recent study by Groundwork Associates Ltd[51] covers some of the above issues in more detail.

12.4 COMPENSATORY MEASURES

Compensatory measures are seen by some operators as a valuable approach[466] but, whilst in general there seems to be some willingness to compensate communities as a whole, there is a reluctance to compensate householders directly[161], eg by money, sound-proofing, an offer to buy. It is recognised that operators inevitably make a contribution to the local economy in the form of rates and the provision of employment. In some cases their workings remove existing dereliction.

Sometimes an operator erects noise bunds almost completely around a single property; this appears to be inconvenient to the operator and detrimental to the householder. A noise/visual fence with some planting might be less intrusive and/or more effective, certainly quieter to erect. Compensatory measures assume greater significance in the 'permanent' or incremental situation.

On the face of it, a programme of compensatory measures would seem to be a better option than a lengthy public inquiry and/or a refusal of planning permission. The principle of compensating neighbouring householders, where an extension to a quarry would degrade their environment, has been accepted. We found that some operators also take this view when new workings are proposed. Double-windows have been provided on a generous

12. Residents

scale for houses adjacent to workings. BCOE have provided high powered TV aerials when high mounds have interfered with reception[324] and they have also provided a mini-bus service where a path was severed[351]. However there are few examples of off-site compensatory measures for individual residents.

Guarantees, eg to provide piped water if wells dry up[31], are more conventional. These conditional offers are likely to be less onerous for an operator than providing facilities before the event. They have the drawback that problems have to occur first, but can ensure that a remedy is provided without too much argument and delay.

Whilst considering use of compensatory measures to minimise adverse consequences for individual householders, guarantee housing mobility and create a good attitude towards an operator, it is appropriate to consider the extension of compensatory measures to the community, ie planning gain.

It is readily accepted that environmental improvements can occur by the removal of existing dereliction, eg due to industry or old colliery workings, in spite of the environmental disadvantages during mining. Another example is the loss of recharge in one section of an aquifer offset by increased protection in another part. A beneficial after-use (examples[315]) is also an opportunity to compensate the community for any dis-benefit it suffers during the operation of a site.

The provision of planning gain is a growing practice when seeking to expand existing sites[264]. Where it is seen by the public as reasonable recompense for tolerating mineral workings, the attitude towards the operation is likely to be much more positive. Naturally, planning gain should not be seen as justification for irreparably damaging the natural environment or a reason to permit wholly unacceptable developments[348]. Indeed, some forms of planning gain, eg a new road, may themselves have adverse effects on a community[110].

We recommend that the Department seek ways to facilitate and encourage the provision of planning gain by operators to compensate individuals and communities for some of the inconvenience, monetary or social costs which they may suffer as the result of mineral workings. At least one MPA[400] expects that operators should help enhance the environment beyond the site boundaries to offset social and environmental costs. It may be that the provisions of the Planning and Compensation Act 1991[468] to permit unilateral undertakings will help.

It is now relatively easy to get financial compensation for loss of property value due to activities of a statutory body, eg Department of Transport building new roads[405]. These provisions are likely to be improved by the Planning and Compensation Act 1991[468]; householders' compensation will be improved and they need not prove that they have been unable to sell.

However, no compensation is available for a 'temporary' inconvenience which may last up to 100 years from surface mineral workings. It has been said[226] that equity demands some

12. Residents

form of similar compensation around opencast workings. One difficulty would be that any scheme of compensation might have to apply to all non-statutory developers. However, as mining can only take place where the minerals occur, there is a possible distinction that can be made in justifying compensation related to surface mineral workings but not for most other forms of development.

Whereas the planning system cannot and probably should not call for compensatory measures, developers/operators should consider providing them in their own self-interest. At least one MPA is in favour[14] saying that, not only should the amenity value of land around settlements be improved by the after-use compared to the original state but, residents should be relocated temporarily or permanently and/or compensated where the effect on their properties is severe. We suggest that operators consider offering to guarantee any loss of property value or to purchase property on demand from a resident within a certain distance of a site boundary who wishes to move before or during operations. This would help to 'remove' sensitive properties from the balance of pros and cons when a planning application is being considered.

To achieve 'sustainability' when using an exhaustible resource (whether it be a mineral or an established landscape) the 'Blueprint for a Green Economy'[335] suggests that there has to be balancing reinvestment in order to avoid compromising the ability of future generations to meet their needs and maintain environmental assets. Although it has been suggested[342] that the UK has considerable aggregate resources and that any marginal annual reduction in these resources is outweighed by the benefits of consumption in terms of the assets created, the vast majority of these resources are either sterilised by environmental constraints or by existing development. Early attempts at a cost-benefit analysis have been made[325] but further research is clearly needed before sustainability can be achieved.

12.5 DURATION OF NUISANCE, INCREMENTAL EFFECTS

The duration of nuisance is a significant factor in determining appropriate levels of control to minimise disturbance, eg 2 years (similar to the life of a construction site) could be considered temporary; 20 years would be effectively permanent. One survey of public reaction[324] found that 9 years, including preparation and restoration of opencast coal sites, was 'much too long' to be considered temporary; 4-5 years had some grudging level of acceptability as temporary.

One basis for judging the significance of duration is the average period during which people occupy a particular house. In the south east during the 1970s and 1980s people moved home about every 7 years.

Opencast coal or sand & gravel working, especially small sites or bigger ones employing progressive restoration, could conceivably work adjacent to a particular group of houses and be restored within 2-3 years; many workings of all types last much longer[463]; a quarry linked with a cement works[428] or a clay pit could be operational for 150 years. There is a clear pressure for progressive restoration and acceptance of that need[338,339,343,351,352].

The Commission on Energy and the Environment[79] referred to the 'temporary' nature of opencast working as a factor to be considered. Whilst disturbance from a single working might be tolerable[351], some areas have been subject to a succession of temporary workings over a period of 40 years[358] and may be 'threatened' with reworking[110]. Whilst any one house may have been directly affected only temporarily, eg by noise, the area has been 'permanently' affected by dust, visual intrusion, traffic, etc. More unsettling in some respects, the area has been in a constant state of change. It is possible that a single integrated and progressively restored site, although larger, would have had an overall shorter working life and produced less impact. It is acknowledged that an application for such a site would probably not have received planning permission.

It is argued[357] that it is better to concentrate disturbance in one area rather than open up new ones. This may be true in terms of actual total exposure to disturbance and adaptation to it, but it is not seen as being fair to local residents.

Minerals Local Plans allow for the formulation of policies relating to the totality of mineral workings in an area. However there is currently no framework for dealing with the incremental effects of a series of successive workings over a period of 20 years or more which may be worse than a single progressively-restored working. It is difficult, and not permissible in law, to judge one planning application on the basis that a second hypothetical one might follow. In relation to opencast coal, the Local Authorities Association[79] have requested that the "totality of the coal deposits in the vicinity" be considered. Research is needed to resolve some of these difficulties.

Further work is needed to define the life of workings which residents consider to be 'temporary'. The effects of 'permanent' workings, or a number of successive workings, including land-fill, ought in our view to be considered along similar lines to those applied to conventional manufacturing industry and subject to correspondingly stricter controls than 'temporary' sites.

Relaxations are often permitted for extremely temporary activities within the life of a working. One MPA has said[351] that it would accept a 2 week breach of a planning condition relating to noise. In some cases certain activities, eg construction of outer faces of bunds or mounds, are exempt from control. We consider that such activities, whilst subject to a relaxation, should nonetheless be controlled. This is in the operator's interest as well, especially when the activity occurs at an early stage of the work and may condition the community's response to later activities.

12. Residents

12.6 HOURS OF WORK

Residents are very sensitive to night-time operations. In addition, early starts often cause significant problems near sites and along access roads (see also Chapters 3 & 4). Roadstone-coating plant may need to run and haulage vehicles accepted on a Sunday to ensure that vehicles can leave the site early Monday morning with material at the right temperature[353].

A sample, but not necessarily a representative one, of planning conditions relating to working hours[1,2,3,17] gave weekday starting times of 06.00-07.30 hrs and finishing times of 16.00-20.00 with some variations for blasting. One survey[324] suggested that residents would find an 08.00 hrs start significantly more acceptable than 07.00 hrs.

There is a need for guidance on transition periods for noise limits in the early morning, as well as in the evening, and how to deal with early starts, including their implications for traffic and minerals operations. Despite this, the need for local variations for specific sites will continue.

12.7 BUFFER ZONES AND LOCAL PLANS

The minimum distance between workings and residents which has been permitted or suggested[1,2,3,16,23,79,91,213,356,400] varies considerably from less than 25-50 m for opencast coal, up to 400 m for sand and gravel and 300-900 m for limestone workings. There is some pressure to increase the minimum distance for opencast coal[14].

Whilst the effectiveness of distance as a means of control varies with topography and local environmental sensitivity and operational factors must be always considered, a little more consistency is desirable in the definition of buffer zones. There is little or no benefit to either operators or residents if the permitted distances are too small. The operators will find that the pressures of cost and expediency encourage them to 'chance it' in tight situations. Residents will find that conditions are often at or over the borderline of acceptability and that the operator is not very responsive to their complaints.

Where blasting takes place the main uncertainty is flyrock. There is no clear picture of what is a safe distance. There appears to be no way in which appropriate planning conditions can be applied to ensure a satisfactory situation that do not conflict with responsibilities of HM Inspectors of Quarries. The situation needs resolution and guidance. Conditions can be set and monitored for noise, ground vibration, peak overpressure and to some extent dust.

Working quarries may be encroached upon by new development, such as housing and golf clubs, taking place nearer to the quarry than may be desirable. Planning conditions, which were set to protect residents in existing development, may not be sufficient to safeguard new neighbours. Conversely, a planning condition may require a standard at the nearest property to the quarry, eg for noise, so new development nearer to it will effectively tighten that standard and restrict methods of working. Operators, not unreasonably, protest about such constraints building up around quarries[428].

One way of protecting both the residents and operator is to designate 'buffer zones' around quarries. This appears to us to be a useful approach in achieving satisfactory relationships between them. Equally it may be appropriate to have a buffer zone around established mineral resources, eg Hertfordshire CC have proposed 60 m[400].

At the moment there is little information or guidance on buffer zones. There would be great benefit from some general guidance on the factors to be taken into account in determining the appropriate width of a buffer zone, and planning conditions which should complement them. The guidance would need to cover at least noise, ground vibration, overpressure, flyrock, dust and possibly visual intrusion.

Formal Minerals Local Plans[a] are recognised as a means of safeguarding mineral resources and controlling the interaction between surface mineral workings and other land uses[414]. Structure Plans deal with these matters in broad strategic terms. Minerals Plans can also contain policies on transport modes, routeing and other environmental factors which will be used to determine planning applications. In principle, there seems to be no insuperable reason why large-scale plans showing detailed buffer zones etc, cannot be incorporated in a Minerals Local Plan, although to date many seem be general statements of policy and strategy rather than specific detail.

Some authorities have relied in the past on informal local plans to record this level of detail, sometimes referred to as 'local quarry plans'[217,454]. These may or may not go through the same public participation process as a formal plan, so may not have the benefit of a full discussion of the issues in a public forum. Given that informal plans are not subject to statutory procedures they may carry less weight for development control purposes.

We can see that there are strong arguments for including strategic land use planning issues, against which mineral working applications will be judged, eg policy on buffer zones, in the formal Minerals Local Plan. On the other hand informal local plans and site-briefs may have a role and be an appropriate way of providing supplementary advice and detail on site-specific matters which are not relevant to the Minerals Local Plan. As such, they could be a useful management tool for the MPAs, District Councils and operators and allow public participation in the wider aspects of site management.

Incidentally the NCC (now English Nature and Countryside Council for Wales) strongly advocated[461] the conservation of Regionally Important Geological Sites (RIGS) which do not enjoy statutory protection. Analogous schemes for the conservation of non-statutory biological sites are often recognised in local plans. RIGS could be recognised in informal local plans.

---

a. This is a Local Subject Plan within the framework of the Structure Plan; CPRE have published a guide[110] which contains a useful summary of the planning system.

12. Residents

12.8 RECOMMENDATIONS FOR FURTHER ADVICE AND RESEARCH

(1) Advice should be provided on the duration of workings and the hours of work in terms of the extent to which acceptable levels of environmental degradation associated with permanent developments may be relaxed.

(2) Encouragement should be given to the use of detailed maps of areas around the more permanent surface mineral workings and significant mineral reserves to keep them apart from sensitive neighbours and vice versa. In this context advice on deciding appropriate buffer distances would be helpful. The actual distances would depend very much on the local circumstances and the nature of the workings.

(3) Off-site compensatory measures should be encouraged to alleviate potential adverse effects, eg provision of double windows, alternative sources of water supply. It is important to encourage operators and MPAs to discuss them prior to a planning application and to published them. Compensatory measures may not always be available to solve a problem directly, so it may be possible to consider an offsetting environmental improvement, eg removing an eyesore instead of providing a bund to limit noise in a garden. In this context consideration should be given to relaxing the present approach to planning gain, subject to the normal reservation that it must not be a excuse to permit 'unacceptable' environmental deterioration.

(4) The significance of the incremental effects of a series of successive workings in the same area should be studied and ways developed to include the consideration of such effects in national and local policies and plans.

(5) There needs to be more understanding of why various sectors of the population object to or complain about surface mineral workings. A study should address the response to the:
- permanent loss of landscape features, habitats, heritage, etc,
- changes of land use for the duration of the workings,
- effects during the life of the workings,
- effects during restoration, including use as land-fill,
- the final restoration state.

Specifically the study should try to determine which features of workings and the approach adopted by the industry are the cause of objections and complaints. The tendency of people to transfer worry about the less tangible effects, eg change to habits of a lifetime, into complaints about physical things like ground vibration and damage should be borne in mind during the study. In view of the complexity of this topic, it would be advisable to commission a preliminary study to design what will be a major exercise.

SUMMARY OF GOOD PRACTICE – RESIDENTS

Good Practice for Mineral Planning Authorities

- make more use of large-scale plans prepared in cooperation with second-tier authorities to help avoid inappropriate development and encroachment in the area around the longer-term workings and significant reserves,
- encourage a dialogue between operators and the community,
- encourage elected members to visit sites before making judgements,
- encourage liaison committees and officers and members to participate in them.

Good Practice for Operators

- be a good neighbour, ie:
    - get to know the neighbours, be concerned about them and try to understand their problems, encourage them to know site personnel, listen as well as talk,
    - provide information as freely as possible, hold open days,
    - create a good impression by running a tidy and efficient site,
- ensure lines of communication, eg:
    - appoint a liaison officer; widely publicise name & telephone number,
    - support a liaison committee,
    - give advance notice and explanation of activities that might cause complaint,
    - keep systematic records of complaints and the remedial actions taken,
    - follow up complaints by personal visits and action,
- ensure that staff are environmentally aware and are trained to cope with the issues,
- do not rely on the letter of the law where there are obvious problems but culpability cannot easily be proved; be prepared to be flexible,
- offer/provide compensatory measures where the impact is excessive or borderline,
- try to cooperate; try to avoid being adversarial.

Good Practice for Neighbours

- get to know the operator:
    - take advantage of any overture,
    - try not to have preconceptions,
    - listen as well as talk,
    - encourage site personnel to visit you at home,
- try to understand the operator's activities and problems:
    - ask for a visit to the site or a similar one,
    - speak to people who have lived in the vicinity of similar workings,
    - take advantage of open days, liaison committees,
- try to be specific when making a complaint, eg date, time, clear description,
- try to cooperate and avoid being adversarial.

CHAPTER 13

EFFECTS ON OTHER SECTORS

13.1 VISITORS, TOURISTS, INVESTORS & HERITAGE

13.1.1 Concerns and Problems

Some interest is shown by tourists in mineral workings but there is concern that, in areas with a reputation for a particularly high environmental quality, tourists will be deterred by minerals operations[16]. Reference is made[353] to visual intrusion and dust spoiling views and photography over many kilometres and coating vegetation. Where it is possible to keep paths open across a site they may not be clear of rubbish and obstructions, or well defined and may be affected by noise, dust and visual intrusion[353]. The landscape setting of historic buildings can be permanently or temporarily adversely affected[243]. Screen planting may itself be inappropriate. Visitors may be distracted by noise and traffic. Historic bridges and green lanes can be damaged by traffic[450].

No literature has been found relating to the effect on potential investors. One opencast coal site[326] was viewed with apprehension by an existing business which feared that it might have to relocate. In the event, there proved to be no problem and the operator was complimented. High-Tech industries, quality offices and food industries are likely to be deterred by the thought of dust; any business will be put-off by traffic congestion.

These deterrents may be feared where it is known that mineral resources exist and may be mined. Some areas are trying to move away from reliance on mining and are encouraging other industries and services. New minerals consents may run counter to such policies.

13.1.2 Good Practice

To counterbalance some of these disadvantages, it would be worthwhile exploring the merits of providing centres to explain the geological, economic and other aspects of workings, making them a positive rather than a negative factor. The National Stone Centre in Derbyshire serves this purpose, BACMI's leaflets 'Why Quarry'[338] designed to inform the public, school children, etc are also helpful.

The scientific and educational value of mineral workings is reflected in the fact that about one-third of all SSSIs scheduled for their geological or geomorphological interest are created by man in the form of quarries, pits, road or railway cuttings[461]. Other than where workings

13. Other Sectors

remove important features, mineral extraction benefits earth science conservation. Useful exposures are created as extraction proceeds, which can aid research, and many active mineral workings are scheduled with the intention of retaining the most scientifically valuable faces as part of the restoration.

To improve the quality of paths that can be retained or provided across sites or diverted around their perimeters, they need to be:
- provided with stiles over walls,
- clearly defined,
- safe when routed between silt ponds and around tips,
- periodically inspected.

For a more detailed discussion of footpath amenity see Chapter 11.

Even though dust may not be considered to be significant in the context of nuisance and damage to neighbours, agriculture or wildlife, consideration should be given to controls on the visual aspects of dust in scenic areas.

13.2 AGRICULTURE & FORESTRY

13.2.1 Concerns and Problems

There are effects due to dust, changes in ground water and drainage which are discussed in Chapters 6, 8 & 9.

The effects of dust are not likely to be more than occasionally significant and will probably emanate from processing plant. There is some evidence that dust, possibly combined with other factors (dewatering, wind), reduces species diversity in adjacent woodland areas and reduces lateral growth of trees. Soiling of agricultural produce and pasture occurs. The latter can affect the uptake of nourishment by cattle. Anecdotally, some precedents for monetary compensation have been mentioned.

Dewatering is often said to be a potential problem but there is little evidence. It can be beneficial in providing better drainage in some cases.

13.2.2 Good Practice

Apart from following the good practice outlined in Chapters 6, 8 & 9 relating to dust, groundwater and surface water, operators should be aware of potential problems and be prepared to take action at the first signs of difficulties.

13.3 ECOSYSTEMS/WILDLIFE

13.3.1 Concerns and Problems

The effects on flora and fauna are 'largely unexplored territory'[461] and more work is needed. There is an EC draft Directive relating to habitats, flora and fauna[99]. Anecdotally, concern for operational difficulties is increasing, however few data have been obtained of real problems. The effects of dust are discussed in Chapter 6, severance in Chapter 11 and agriculture and forestry above.

Minerals can only be mined where they occur and nature can only be conserved where it occurs[163]. Concern is expressed[382] for affects on neighbouring ecosystems by:
- drainage affecting nearby ponds and wildlife,
- the removal of feeding areas,
- the emissions of dust, fumes and noise.

Dust is unlikely to be a problem except in very special circumstances. In one case it increased the species diversity of lichens[296]. Animal trails are likely to be affected but no specific data have been found.

Dewatering can affect adjacent SSSIs especially in bog or raised mire areas[15]. Areas of unworked peat may be significantly affected[455] not only by dewatering but also by the method of working which involves stripping of vegetation over a wide area causing slumping and oxidation. Recharging by pumping to avoid the problems of dewatering sensitive ecological sites, eg peat bogs, may not be appropriate because the water quality will be different from rain water. It is said to be impossible to dig or drain part of a bog without affecting, albeit very slowly, the remainder. Slow but steady impact occurs in terms of loss of species adapted to the natural stable bog water table[402].

Potential problems exist and concern is expressed particularly for the rarer animal species[449] in relation to:
- noise and disturbance, particularly during the breeding season,
- water pollution,
- a general increase in the presence of people.

Surface mineral workings sometimes create valuable habitats for wildlife and their subsequent destruction, eg of birds nests, causes problems for operators[344]. Some sites have reported birds nesting whilst work is in progress and having requirements imposed/agreed to protect the nests until the young birds are fledged and/or for future seasons. Examples quoted include, a cliff face kept for Kestrels[13] and sand face for Sand Martins. Birds may however be unaffected by nearby operations; one report[13] was of the operations having no effect except that the birds cocked their heads at reversing bleeps. Among the shy species that might be disturbed are Heron, Green and Greater Spotted Woodpecker[472].

13. Other Sectors

13.2.2 Good Practice

Apart from following the good practice outlined above in Chapters 6, 8 & 9 relating to dust, groundwater and surface water, consideration should be given to:
- having a policy to minimise the conflict with nature[412],
- carrying out an ecological baseline survey and continuing to monitor during the life of the workings; especial attention should be given to the possible presence of protected and rare species, eg badgers[385],
- preserving valuable habitats, eg green lanes[477], avoiding disturbing badger setts[413], retaining foraging grounds allowing badgers to co-exist with the workings; failing this relocating the setts,
- maintaining some space adjacent to woodland habitats especially where wildlife is known to feed[385],
- leaving margins around or along trees and hedges, eg 4 m for hedges, 5 m beyond the spread of trees for hedges with trees, 10 m for trees and 15 m for woodlands[384,383],
- leaving margins for ditches[362],
- using impermeable membranes to prevent loss of water from ponds[382],
- transplanting hedges[362] but not during the nesting seasons[384],
- transferring habitats temporarily or permanently[188,245],
- stopping work in the vicinity of nests during the breeding season of protected birds, eg Sand Martin and Little Ringed Plover[344],
- ensuring that at least part of a suitable habitat is always available for any rare species[383],
- phasing workings adjacent to woodlands[382] and progressively working and restoring sites[384] to give the ecosystem more chance to survive and to recover naturally.

In general, operators should be aware of potential problems and be prepared to take action at the first signs of difficulties.

Compensatory measures to improve and manage adjacent areas, possibly by transplanting a habitat, are sometimes advocated and attempted[375,432,481,482]. Where an excavation is to be left unfilled and either exposed or flooded, it has been suggested that leaving it untidy and subject to fluctuating water levels encourages interesting species of flora and fauna[153].

13.4 ARCHAEOLOGY

13.4.1 Concerns and Problems

Problems are usually due to insufficient communication and cooperation between operators and archaeologists.

When the operational programme includes progressive working and seasonal soil stripping, 'watching briefs' are sometimes maintained in the expectation that surveys can be done more readily once the top soil is removed. Unfortunately the stripping of topsoil by a contractor, especially without archaeological supervision, is often unsatisfactory[437]. The remains are often damaged because the soil is stripped in wet weather or are destroyed because the soil, including the archaeological remains, is stripped down to clean gravel.

The importance of individual finds during mineral working is rarely of great significance. Current interest is in the environment and context of settlements rather than in artefacts. In the event of rare finds there seems to be no alternative to relying upon the goodwill of the operator to alert archaeological interests.

Some remains are covered by alluvium. They may be pre-historic remains washed down river and covered by later sediments. Such sites are of particular interest. These are very difficult to identify by the conventional survey techniques of aerial photography and trenching. It is hoped that new survey techniques will be available shortly.

Serious problems occur in wet areas where dewatering of a minerals site can dry out adjacent areas containing artefacts which may depend upon submersion for preservation[241,-244,246]. It is not always easy to predict the extent of dewatering or its effects[436]. Waterlogged archaeological sites occur in peat and gravels of major river valleys[450]. The soil in intermittently wet areas may have sufficient clay in it to retain enough moisture to preserve the remains. If the regime is even slightly altered by mineral working, the change may be enough to initiate rapid decay.

13.4.2 Good Practice

The CBI code of practice for minerals operators[10] has recently been reviewed. Some MPAs require access for observation and appropriate response to finds, eg allowing local museums to retain them[400,455].

PPG16[123,130,469] says that it is appropriate for MPAs to expect that a developer will agree arrangements for a survey but not necessarily to pay for it. It is suggested that developers should consider insuring themselves against unexpected finds. CBI[10] supports the concept of operators offering financial or practical assistance.

The need for surveys, where there is no direct evidence, can be inferred from experience of adjacent areas. The minerals industry however probably needs reassurance that surveys are not sought unnecessarily. Equally it should consider employing its own consultants when in doubt[10,455].

Efficient pre-operational surveys will identify most significant archaeological interests and decrease the likelihood of problems during working of a site. The choice, before operations commence, is between preservation and destruction by investigation. In the case of sand

13. Other Sectors

and gravel workings, archaeological interests can be relatively cheaply preserved by leaving 'islands' undisturbed.

To be archaeologically effective, soil stripping has to be done in two stages, viz:
- grass and immediate topsoil only to allow the archaeological survey to be undertaken,
- remaining soil.

Such careful stripping is expensive and watching briefs are going out of fashion. At least one operator however permits the archaeologists to supervise the stripping of top-soil and encourages cooperation.

Control of waterlogged remains[243,450] is possible by:
- leaving an undisturbed safety/buffer zone around the remains,
- recharging the area and continuously monitoring the water level to ensure that it does not fall too low; this can be acceptable if the water level will be high enough when dewatering stops,
- creating a bund around the area and filling it with water to ensure that the remains are kept submerged.

13.5 RECOMMENDATIONS FOR FURTHER ADVICE AND RESEARCH

(1) There is considerable uncertainty about the significance of effects on neighbouring ecosystems. More practical guidance would be useful on the recognition of sensitive features and about reasonable precautions that should be taken at different levels of sensitivity.

Research may be necessary to provide the basis for this advice. From the little literature available, it would seem that this would be a long-term and extensive programme outside the remit of the Department's Minerals and Land Reclamation Division.

One practical shorter term option might be an in-depth review of available information of effects upon ecosystems of activities and disturbances generally which are similar to those arising from mineral workings. See also Section 9.5-3 for the suggestion of research relating to the irrigation of areas adjacent to workings.

SUMMARY OF GOOD PRACTICE – OTHER SECTORS

Good Practice for Mineral Planning Authorities

Consider the need to encourage or agree or specify planning conditions relating to:
- paths and facilities to encourage the tourist potential of workings,
- the control of dust in scenic areas,
- baseline studies of the ecosystem,
- buffer zones between workings and sensitive habitats, etc,
- progressive working and restoration to ensure continuity of habitats,
- liaison with archaeological interests,
- the provision of interpretive centres.

Good Practice for Operators

- consider opportunities to make workings positive features for tourism and scientific interest,
- provide educational opportunities and material,
- consider the provision of interesting paths across safe parts of their workings,
- recognise that dust may be a problem in scenic areas even though uninhabited,
- be aware of ecological issues, if necessary employing a consultant to carry out a baseline survey prior to finalising any plan,
- be on the look out for early signs of distress in the ecosystem or agriculture and be prepared to act before significant problems occur,
- accept the need for buffer zones around sensitive habitats, etc,
- progressively work and restore wherever possible to minimise the risk of permanent damage to the ecosystem and to maximise the speed at which it will recover,
- consult and be flexible in dealing with archaeological interests and where in doubt commission a survey.

CHAPTER 14

DISCUSSION OF RELATED MATTERS

14.1 PLANNING TECHNIQUES

14.1.1 Environmental Assessment

As a result of the EC Directive on Environmental Assessment[63] and corresponding UK regulations and procedures[100,122,192], the planning of surface mineral workings has been forced to embrace more systematic reviews of the environmental effects. It is clear from those Environmental Statements we have seen that the quality varies widely. At one extreme they are superficial reviews of the issues with an acknowledgement that there are a few potential problems which the operator will overcome or avoid in unspecified ways. At the other extreme they include detailed analyses of impacts, undertakings to employ specific mitigating measures and objective statements of the residual impacts against which subsequent performance can be monitored and compared.

There is a clear opportunity in the procedures of environmental assessment, including the level of consultation that is involved, for greater understanding to be reached between the operators, their neighbours and the regulatory authorities. There is also the opportunity to use the results of assessments as a basis for monitoring of the actual environmental effects and to determine whether planning decisions were well-founded. As yet it is too early to tell whether this opportunity will be taken.

There are situations where there needs to be a trade-off between a variety of environmental factors, eg in the use of mounds – between visual amenity, noise reduction by screening, noise created during construction and surface water run-off. In one case a dirt road (rather than one surfaced with macadam) was specified to minimise the scar on the side of a hill; in practice it suffered from considerable run-off and erosion. There is a need for an approach equivalent to Integrated Pollution Control; it is of course recognised that HMIP are not responsible for surface mineral workings. Environmental assessment should help to resolve such issues and improve the broader consideration of the environment.

14.1.2 Computing Technology

Across the range of technologies associated with the surface minerals industry (reserves, extraction planning, blasting design, vibration control, visualisation, noise, haulage) computerisation is playing an increasing role[198]. Its use is increasing the extent to which it

14. Discussion

is possible to integrate and optimise decisions. The use of computer systems imposes a discipline and helps to level-up the worst practice without perhaps significantly improving the best. Computer modelling can help to optimise the balance of environmental costs and benefits[392].

At present there is a problem with data acquisition and input, which may be practical and economic for a large site but uneconomic for small sites and companies. The situation is improving for two reasons:
- the restructuring of the industry into larger units increasingly facilitates the use of such techniques,
- progress in the techniques of data acquisition and computer software is steadily reducing the size of operation for which the techniques are economically viable. Specifically, there are advances in mine planning[28,52], blasting and monitoring[98,103], visual assessment and planning[41].

Whilst there are examples of excellently planned sites, it is unclear to what extent the industry as a whole plans, with or without the above techniques, to achieve good environmental operational conditions. There is increasing use of palliative and cosmetic measures but changes to the basic design of mines to minimise problems is not obvious, except for restoration.

14.2 TRENDS

Changes are taking place in planning, mining practice and the economy; some trends are emerging[52,67,70,171,212,484,488]. Economic and planning pressures are tending to create a demand for larger quarries which can take advantage of one or more of the following opportunities to:
- justify more detailed planning, monitoring and control,
- justify more effective preplanting,
- be remote from habitation and tourists,
- encourage other industries and socio-economic development,
- justify a railhead or jetty and permit the selection of non-sensitive locations for the transfer of material to lorries for delivery,
- allow mining costs to be optimised without too many constraints,
- permit working below the water table if necessary,
- possibly justify the use of in-pit conveyors and crushers.

Other reported trends are:
- a smaller proportion of rural residents making their living locally,
- more urbanisation creating a greater need for rural and aesthetic recreation,
- increasing reaction to proposals for new mineral working,
- demands for higher environmental standards.

These trends, together with the increasing demand for minerals, create the now classic pincers of growth and concern for the environment with the mineral operators between them. The situation is becoming progressively exacerbated as workings increasingly impinge on sensitive areas as the more suitable locations are worked out. The situation could be eased by the better use of wastes and the recycling of used aggregates[125].

There are pressures for increased working depths which may affect the water regime and quality. Some mineral planning authorities[463] have considered the value of restricting the area devoted to quarries by permitting sub-water table working but were unsure of the consequences. NRA however are trying to avoid workings which might prejudice aquifers. It may be useful to study, at a strategic level, the advantages and disadvantages of deeper workings compared to a larger number of shallower workings, including the effects of sub-water table working.

Advice is desirable on any environmental issues where standards should be expected to be common to various parts of England and Wales. Some sections of the industry seem to aim for common standards regardless of the area in which they are working. We feel however that areas vary in their environmental sensitivity and that MPAs can reasonably expect standards for noise, dust, vibration, landscaping and traffic to differ according to the nature of the site and its surroundings.

The solutions for most environmental problems are available but their use may be constrained by the operator's unwillingness or inability to pay for their use[212]. There needs to be a combination of proper planning procedures to determine and evaluate in a community context all the factors involved, with a realistic awareness of the operator's ability to pay.

There are trends in mining practice which are helping to solve environmental problems, eg:
- the greater use of the quieter electrically-powered plant,
- the development of conveyors which can 'climb' steep gradients (minimising the need for internal haul-roads),
- the recognition of the need for greater care in surveying of faces and drilling (for reasons of health and safety),
- the availability of more sophisticated techniques for blast design, blasting and measurement of performance.

Nonetheless it is claimed that the costs of production will generally increase as greater degrees of environmental control become necessary[340] and that the resulting increase in costs would be effectively an environmental tax.

14. Discussion

14.3 MONITORING & ENFORCEMENT

14.3.1 Monitoring

There is a need for baseline monitoring prior to a planning application, permission and development, eg of noise, groundwater, surface water courses, ecosystem; the effective monitoring period varies. For noise the period can be relatively short, ecosystems however will need observation preferably over the seasons, and it may be necessary to gather data over several years in order to understand the water regime.

Such baseline monitoring is also essential for the preparation of a meaningful Environmental Statement, which can be used as a basis for subsequent monitoring to measure or assess noise, dust, traffic, landscaping, etc to provide grounds for any enforcement activities. More importantly perhaps, operators have the opportunity to monitor/audit their own performance and to identify trends which may lead to problems and/or breaches of planning conditions.

The ease with which this can be done varies. Within the working, the precise location of mobile noise sources may change daily, hourly or even more quickly, in contrast to a working face which will move more slowly. The initial breach of a planning condition may have only a small effect, but by the time the effect is significant it may be incapable of remedy. Once people have been sensitised, upset and possibly moved to submit even isolated complaints, it may take a great deal of effort to change their adverse attitude.

It has been suggested[217,232] and implemented that MPA personnel visit each site at least twice a year; a checklist derived from these suggestions is to report changes in:
- limits of the excavation,
- the landscape,
- amenity,
- drainage,
- water supply,
- traffic,
- blasting methods,
- records of blasting, noise, etc.

It has been suggested[396] that it eases any access problems for regulatory authorities and operators if monitoring of some factors can be at the same locations, eg site noise, blasting effects, dust.

Operator should consider, even if not required to do so by their permissions, monitoring their activities and the effects of them.

14. Discussion

A monitoring programme could be on the following lines[383]:

Daily:
- condition of internal and public roads,
- dust emissions, accumulation of dust on internal roads during dry periods to identify the need for spraying,
- the use and effectiveness of wheel and chassis cleaning facilities,
- water level in the nearby ponds, during the summer,

Weekly:
- condition of drains,
- quality of surplus water discharged,
- irrigation requirements for agricultural land and for landscape planting (in the summer only),

Monthly:
- dust deposition at site boundary,
- condition of footpaths, stiles, security, fences, signposts[384],

Twice Yearly:
- levels in licensed ground water abstraction points,
- water chemistry in ponds, oxygen, pH, conductivity,

Annually:
- the need to replace failed planting,
- condition of stored soils.

14.3.2 Enforcement

There is some suggestion that, desirable as they may be, planning conditions often remain unmonitored and unenforced. It is suggested that this is due to a lack of MPA, NRA and LA staff resources and because of the time it takes to prosecute.

The enforcement system has been variously described as:
- tending to be the "Cinderella of the planning system"[294],
- having in terms of landscaping "examples where schemes had not been carried out even after a lapse of several years ... no enforcement action"[286],
- having insufficient resources and tending to crisis management; monitoring and enforcement based upon public observation and complaint[443],
- "notoriously cumbersome, difficult and longwinded"[22],
- needing "more effective sanctions"[429],
- most complex, time consuming and uncertain"[433],
- "can be slow and procedures cumbersome"[435],
- "a stop notice risks heavy compensation"[440],
- "inadequate, ...with the delay arising from the procedural requirements..."[440].

14. Discussion

There is a need to strengthen and simplify the control procedures[463]. It has been suggested that the procedure and more extensive powers available to control waste disposal sites under the Control of Pollution Act 1974 (now covered by Part II of the Environmental Protection Act 1990) might be effectively applied to mineral workings, especially for 'minor' offenses[435].

A Department sponsored study and report on planning enforcement[328,349] confirmed that the planning system, and the scope given to operators to create delay, is widely criticised. Unless a stop notice is issued, it can be years before an operator is forced to act. Stop notices are rare because of the fear, not necessarily justified, that a successful appeal will lead to compensation. The Department has accepted many of the report's recommendations and the Planning and Compensation Act 1991[468] provides for:
- 'planning contravention' notices,
- 'breach of condition' notices,
- non-compliance with these notices to be a summary offence,
- injunctions against actual or threatened breaches of conditions,
- improved procedures.

Bonding arrangements in the UK are voluntary and cannot be required by MPAs. BCOE have successfully operated bonding arrangements with their opencast contractors for many years. Arrangements are also made on an individual voluntary basis for restoration and for landfill gas and leachate control[440]. Bonding could be considered, rather than relying on prosecution, for pre-planting, compensation if work overruns, when operators delay compliance with planning conditions[17] and when water is polluted. The potential uses of bonds are the subject of a current study[127].

It is said[26] that some planning conditions are unnecessary because of existing laws and regulations. For example, the Construction and Use Regulations 1986 for road vehicles, which make it an offence to shed any loads on the highways, render any planning condition to that effect unnecessary. It is however difficult for the Police to enforce the Regulations and much easier for an MPA to insist on the sheeting of vehicles; also the enforcement of legislation is reactive after the deed but planning conditions can be preventive.

There are desirable objectives, eg the routeing of lorries, which are not suitable for planning conditions or Section 106 agreements. There appears to be a need to frame 'standard' forms for unilateral legally binding agreements to cover such matters.

There is a multitude of regulatory agencies with various responsibilities relating to surface mineral workings; they include:- HSE (health and safety), MPAs/LAs (minerals planning, land use, landscape, water, highways, environmental health), NRA (water), Police (traffic). Unfortunately, the priorities of these agencies are not always the same as those of the planning authority, so there is potential for conflict[145]. The recent Environmental White Paper[146] in paragraph 6.39 says that LAs 'must consider all effects' and planning permission 'should not be granted if that might expose people to danger'. We feel that the various agencies do not always form a coherent and well-balanced view and there should be more focused cooperation between them.

Bearing in mind the lack of resources for frequent site visits, we consider that there is a need to make the best use of them. Collaboration and cross-disciplinary working should be encouraged. For example, the possibility of nominating an inspector/observer from one of these agencies should be examined. Their reports could be issued to and acted upon by some or all of the agencies with a responsibility relevant to mineral workings.

One operator is reported as saying[401] that the regulatory pressures on the industry are relatively lax, they have to be their own policeman and 'the big stick' is that the operator's track record is increasingly taken into account in planning decisions on new workings.

14.3.3 Self-Auditing

Mineral operators would probably find it advantageous to have their own in-house standards which cover all the relevant factors and interpret the regulatory and their own requirements into practical terms, eg sheeting of lorries, conditions for safe blasting. In addition, they would surely find it advantageous to monitor compliance for themselves and not wait for complaints from the public or the regulatory agencies. BACMI[125,170] is working on an environmental code of practice for its members.

It has been suggested[321] that it is better for an operator to take positive initiatives about 'mistakes' and ask local authorities for help, rather than to wait until a breach is discovered. This needs self-auditing against predetermined performance requirements. A recent study[51] of the minerals industry recommends that each operator should have an environmental policy and a management system which includes the auditing of activities and also that the industry should collectively produce a series of codes of practice, eg on traffic, dust and community links.

These proposals take place against an ever growing impetus for environmental auditing and its application throughout commerce and industry internationally[131], eg:
- International Chamber of Commerce[139],
- draft EC Regulation[141] encouraging organisations to register for a scheme which would require an annual statement and audit; this has been subject to consultation[194,196] and widely discussed[129,195],
- draft British Standard[140,193] for a system of auditing and registration akin to Quality Assurance,
- Chartered Association of Certified Accountants[189].

Some operators are already doing this[401] without necessarily using the term audit. BCOE are establishing environmental audits by teams not involved in the management of the site[456]. In their existing arrangements whereby work is subcontracted, planning conditions are made part of the sub-contracts and are monitored and enforced by BCOE. This approach has made life easier for regulatory authorities. The results of audits not only check compliance but also provide feed-back for future planning and design[392].

14. Discussion

It has been suggested by industry in general, not just by the minerals sector, that audits should be voluntary and unpublished as this encourages more open internal dialogue between assessors and the assessed.

Many of the regulatory bodies are understaffed and would probably willingly allow greater flexibility to operators known to be efficiently policing themselves. Operators could enter into agreements to have a Quality Assurance system which includes internal audits (as do some waste/landfill contractors) or to commission periodic environmental audits by an independent organisation.

Some local authorities are sceptical about self-regulation but accept that voluntary codes of practice have value and can complement the planning authorities' statutory powers[463].

14.4 RECOMMENDATIONS FOR FURTHER ADVICE AND RESEARCH

(1) Advice to clarify the desirability or otherwise of having common standards regardless of location would be helpful, particularly in view of the tendency to quote precedents at public inquiries.

(2) The Department should encourage the regulatory authorities to cooperate in considering, approving and the monitoring the performance of a surface mineral working to maximise their collective effectiveness. This would be aided by simpler and more streamlined enforcement procedures. A study of the ways in which this could be done may be necessary.

(3) Guidance would be useful on forms of legally binding agreements to cover matters not properly the subject of Section 106 agreements. The Planning and Compensation Act 1991[468] provides for developers to make unilateral undertakings.

(4) Self-auditing and self-regulation by operators should be encouraged. MPAs and other regulatory authorities should be advised as to the extent that the existence of such procedures should influence their reaction to planning applications and their own enforcement activities.

(5) A strategic study of the pros and cons of deeper, in return for perhaps fewer quarries, especially where this involves working below the water table, is tentatively suggested.

SUMMARY OF GOOD PRACTICE – GENERAL

Good Practice for Mineral Planning Authorities

Good practice, resources permitting, should include:

- periodic monitoring of all sites,
- cooperation with other regulatory authorities to share the load,
- systematic enforcement,
- encouraging self-auditing and self-regulation by the operators.

Good Practice for Operators

Operators should consider:

- the use of 'modern' techniques for the planning and optimisation of their workings to maximise the economic benefits and to minimise the environmental effects,
- periodic monitoring of compliance with all planning conditions, NRA consents, licences and other factors, not subject to planning conditions, that might lead to adverse effects,
- liaison with MPAs and other regulatory authorities immediately a difficulty is encountered,
- the introduction and use of self-auditing and self-regulating procedures.

CHAPTER 15

SUMMARY OF RECOMMENDATIONS

FOR FURTHER ADVICE AND RESEARCH

15.1 GENERAL

Much work can be done to ameliorate the environmental effects of surface mineral working by implementing many of the techniques and practices which are currently available and discussed in this report. There are numerous areas where further advice on policy and technical issues is required by MPAs and operators. Better information will extend the range of options, improve decision making and help ameliorate the effects.

The recommendations are divided into those where:
- advice can be issued without significant further research,
- advice on criteria and standards is required but must be preceded by significant research,
- codes of practice should be developed,
- other advice based on research is needed; in this case an indication of the time scale is given.

Whilst these recommendations are made to the DOE, we do not consider that the responsibility to fund and undertake this research lies solely with the Department. Many of the recommendations, particularly those for codes of practice, monitoring studies and research into detailed techniques, could be initiated by mineral operators, trade associations, local authorities or their associations and other Government agencies.

Throughout cross-references [in square brackets] are given to fuller discussion of the recommendations.

15.2 ADVICE

Advice, which could be given with limited further research, would be helpful on:
- greater cooperation between regulatory bodies [14.4/2],
- the use of detailed large-scale plans for areas around longer-term workings and significant mineral reserves [12.8/2],
- suitable distances between workings and sensitive land uses where flyrock is a possibility (with the Health and Safety Executive) [4.5/2],

15. Recommendations

- avoiding the use of surface detonating cord (with Health and Safety Executive) [4.5/4],
- the use of structural surveys prior to blasting to reduce arguments about damage [4.5/5],
- ways to facilitate and encourage pre-planting on- and off-site [7.5/3],
- the encouragement of more monitoring of groundwater and surface water together with any effects, such as derogation, changes in drainage, flora and fauna [8.5/2].
- self-regulation by operators, its encouragement and its implications for Mineral Planning Authorities, etc [14.5/4],
- the provision of off-site compensatory measures [12.8/3],
- desirability or otherwise of common standards compared to site-specific ones [14.4/1].

15.3 CRITERIA AND STANDARDS

Advice on criteria and standards, which would require significant preparatory research, would be helpful on:
- the degree to which environmental requirements associated with permanent developments can be scaled according to the duration of workings, eg 2 to 100 years and further relaxed for individual activities down to a few days [5.5/1, 12.8/1],
- the significance of incremental effects and how to reflect them in national and local policies and plans [12.8/4],
- understanding blasting nuisance from the combination of overpressure and ground vibration and guidance on acceptable levels is strongly recommended [4.5/1]; the research necessary for well/founded advice is likely to take a several years and some interim outline guidance would be useful,
- acceptable noise levels, particularly for the night and dawn periods [5.5/1],
- acceptable levels of noise, ground vibration and overpressure from surface mineral workings and guidance on the course of action if these levels are exceeded at proposed development sites adjacent to existing workings; this should bear in mind the duration of working [4.5/3, 5.5/3],
- criteria and limiting levels for annoyance and nuisance due to dust levels and the means of measurement [6.5/2]; a social survey of public response to various levels/rates of dust deposition should be included in the work,
- requirements and guidelines for groundwater (National Rivers Authority) [8.5/1],
- appropriate limits for discharges to surface water courses, for alterations to surface water courses and effects on the environment of river corridors (National Rivers Authority) [9.5/1],

15.4 CODES OF PRACTICE

An important medium-term aim should be to develop codes of practice for each main type of mineral working covering all the relevant environmental concerns [2.4/2]. This and most of the following recommendations will require significant research.

Specific subjects for technical advice on means of control, which could be in the form of codes of practice, are:
- the effectiveness of the various types of vehicle cleaning facilities [3.5/3],
- the provision of sheeting facilities; encouragement of the more convenient semi-automatic forms of sheeting [3.5/1],
- the quieter forms of reversing alarms (with Health and Safety Executive) [5.5/2],
- up to date noise data on plant and equipment (liaise with Construction Industry Research and Information Association) [5.5/4],
- dust control methods, design of stockpiles, use of additives, mists, etc is badly needed [6.5/1],
- the design of settling lagoons (National Rivers Authority) [9.5/2],
- footpaths and animal runs [11.5/1].
- the recognition of sensitive features in the neighbouring ecosystem (English Nature and the Countryside Council for Wales) [13.5/1].

15.5 OTHER RESEARCH

Research beyond that necessary for the development of criteria and codes of practice and which is desirable but less pressing should include:

short- to medium-term:
- the choice between different types of planning condition, ie performance levels, control measures and a combination with possibly a 'trigger' level [2.4/3, 6.5/3],
- how advantage can be taken of technological progress during the life of long-term workings to limit environmental effects [2.4/1],
- the development of lorries/tippers which are quieter when empty, especially those avoiding body-slap [3.5/2, 5.5/5],
- situations in which other measures would be more appropriate than the use of bunds to protect individual or small groups of houses and footpaths [7.5/1],
- the design of a study to identify why the public registers objections and complaints about surface mineral workings; this should take account of attitudes and that people sometimes transfer objections to the more tangible effects [12.8/5],
- to design a programme to study the effects on flora and fauna, including agriculture and woodlands, due to changes to groundwater and surface water regimes [8.5/3, 9.5/3],

15. Recommendations

medium- to long-term:
- forms of legally binding agreements or undertakings to cover matters not properly the subject of Section 106 agreements [3.5/4, 14.4/3]
- the environmental costs and benefits of deeper, in return for perhaps fewer quarries, is tentatively suggested [8.5/4, 14.4/5],
- the environmental costs and benefits of dry-working of sand and gravel [8.5/5],
- the sequencing of quarry working to facilitate progressive restoration [7.5/2],
- the production of less waste [10.5/1],

longer term:
- the assessment of the environmental capacity of individual rural roads [3.5/5].

CHAPTER 16

ACKNOWLEDGEMENTS

16.1 GENERAL AND IMMEDIATE TEAM

Roy Waller Associates Ltd gratefully acknowledges the help provided by many people and organisations. They gave freely of their time, knowledge and resources; any errors are ours and not of those who helped and whose advice may not have been taken. The majority are listed below; we apologise for any omissions.

We would like to thank the staff of Minerals Division, DOE for their help and support, particularly Robin Mabey, Andrew Routh, Sharon Cosgrove (who was the Nominated Officer for the project), Ann Ward, Peter Crawshaw, Alastair Bishop and David Crossley.

We would also like to thank the other members of the Steering Group for their expertise and assistance during the course of the project:-

J Adams	British Coal Opencast Executive
E J Allett	British Coal for CBI (initial period only)
J Bailey	Surrey County Council for LAAJMARG/ACC
M Bellingham	Blue Circle Industries plc for CBI
S J Cribb	Minerals Industry Research Organisation
M Davies	Brett Gravel for Sand and Gravel Association
K Duff / C Stevens	English Nature (formerly NCC)
M G Tole	Walsall MBC for LAAJMARG, CPO Society Committee 3
D T Pollock	British Aggregates and Construction Materials Industries
G Hughes	Steetley Quarries for National Federation of Clay Industries
E Darlow	Health and Safety Executive
T N Reeder	National Rivers Authority for WAA
J Sargent	Building Research Establishment
	– (W Utley until his death)
I Thomas	Association of Metropolitan Authorities
A Pursell/J Owen	Association of District Authorities

We would like to thank those individuals and companies who joined with Roy Waller in the project team for their work, advice and support:-

Alastair Baillie	Alastair Baillie Associates
John Bailey	Bailey Mining Consultants Ltd
Chris Down	
Ewart Parkinson	
Keith Jones	Land Capability Consultants – during the first phase of the work,
Mike Chamley	Hunting Land & Environment Ltd, formerly Land Capability Consultants
Jeff Stevenson	initially Nicholas Pearson Associates, latterly RPS Clouston

16. Acknowledgements

16.2 RESPONSES

Responses to public and private invitations to comment were received from:-

Association of District Councils
Avon, Gloucester & Somerset, Environment Monitoring Committee
Blue Circle Industries plc
Brett Gravel Ltd on behalf of Sand and Gravel Association
British Aggregate Construction Materials Industries (BACMI)
British Coal Opencast Executive (BCOE)
Buckinghamshire CC
AF Budge (Mining) Ltd
Cleveland CC
Clouston
Cornwall CC
Community Technical Aid, Northern Ireland
Dyfed CC
Dyfed CC (National Park Office)
English Heritage
Essex CC Archaeological Unit
Golder Associates (UK) Ltd
Hampshire CC
Health and Safety Executive (HSE)
Humberside CC
Minerals Industry Research Organisation
National Federation of Clay Industries Ltd
National Rivers Authority
Nature Conservancy Council
Oxford Archaeological Association Ltd
Oxford Archaeological Unit
Peak National Park
Redbridge, LB of
Rock Environmental
RPS Clouston
Royal Society for the Protection of Birds
Sheffield City C
Somerset CC
South Bucks DC
South Glamorgan CC
Strathclyde Regional Council
Suffolk CC
Surrey CC
Tilcon
Wallsall MBC
Warren Spring Laboratory
Wessex Archaeology, The Trust for
West Glamorgan CC
West Sussex CC
R A Young Mining Ltd

16. Acknowledgements

16.3 MEETINGS

Meetings were held with:-

    BCOE
    Blue Circle Industries plc
    County Archaeologists, Association of
    Cumbria CC
    English Heritage
    Grimley JR Eve
    Hertfordshire Groundwork Trust
    Laporte Minerals
    Leicestershire CC
    Mendip DC
    Nobel's Explosive Co Ltd
    Peak District National Park
    RMC
    Rock Environmental
    Rotherham BC
    Somerset CC
    South Glamorgan CC
    WS Atkins

16.4 TELEPHONE DISCUSSIONS

Telephone discussions were held with:-

    Allerdale DC
    Arup Acoustics
    Coal Contractors Ltd
    Essex Archaeological Unit
    Fenland Archaeological Trust
    Hampshire CC
    HM Inspectorate of Mines and Quarries
    Lichfield DC
    London Brick Co Ltd
    Minerals Industry Research Organisation
    National Rivers Authority (several regions)
    North West Leicestershire DC
    Oxford Archaeological Unit
    Staffordshire CC
    Surrey CC
    Sutton LBC
    Warren Spring Laboratory
    Watts Blake & Bearne Ltd

16. Acknowledgements

16.5 SITES VISITED

The sites visited were:-

Sand and gravel:

    Kingsmead (processing only) – RMC
    New Platt Wood – Hepworth Chemicals
    Alrewas – Redlands Aggregates Ltd
    Shepperton – Streeters Ltd

Hardrock:

    Batts Coombe (limestone) – ARC
    Moons Hill Quarry (basalt) – John Wainwright Ltd
    Mount Sorrel (granite) – Redlands Aggregates Ltd
    Sharpstone Quarry – Tarmac Roadstone (Western) Ltd
    Torr Works (limestone) – Foster Yeoman Ltd
    Watley Quarry (limestone) – ARC

External tour in Peak District:

    Tunstead
    Old Moor
    Doveholes
    Eldon Hill
    Hope Valley

Coal:

    Pit House West – BCOE
    Potato Pot – BCOE
    Lounge – BCOE

Peat:

    External tour of various workings in the Somerset Levels

Clay:

    South Holmwood – Redland Bricks Ltd

CHAPTER 17

GLOSSARY OF ABBREVIATIONS AND TERMS

17.1 ABBREVIATIONS

BACMI	British Aggregates and Construction Materials Industries
BATNEEC	Best Available Technology not Entailing Excessive Cost
BCOE	British Coal Opencast Executive
BPEO	Best Practicable Environmental Option
BPM	Best Practicable Means
BS	British Standard
BSI	British Standards Institution
CBI	Confederation of British Industry
CC	County Council
CIRIA	Construction Industry Research and Information Association
COPA	Control of Pollution Act 1974
DC	District Council
DOE	Department of the Environment
EC	European (Economic) Community
EPA	Environmental Protection Act 1990
HGV	Heavy Goods Vehicle
HMIP	HM Inspectorate of Pollution
HSE	Health and Safety Executive
IPC	Integrated Pollution Control
LA	Local Authority
MBC	Metropolitan Borough Council
MIC	Maximum Instantaneous Charge (in blasting)
MPA	Mineral Planning Authority (CC or MBC)
MPG	Mineral Planning Guidance note (published by Department of the Environment/Welsh Office)
NCC	Nature Conservancy Council (now replaced by English Nature and The Countryside Council for Wales)
NFCI	National Federation of Clay Industries
NIMBY	Not in my back yard
NRA	National Rivers Authority

17. Glossary

PPG	Planning Policy Guidance note (published by the Department of the Environment/Welsh Office)
SAGA	Sand and Gravel Association
SSSI	Site of Special Scientific Interest (designated by NCC)
USBM	United States Bureau of Mines

17.2 TERMS

Circular	Note setting out Governmental advice to LAs.
Community	European Economic Community.
Compensatory Measures	Measures taken, usually outside a site, to reduce the impact of a working on neighbours; this can be direct in the form of sound-proofing or indirect, such as the gift of a community amenity such as a recreation ground.
Decking of Charges	The division of the explosive in a single drill hole into two or more separately detonated charges to reduce the MIC.
Department	Department of the Environment.
Derogation	The reduction of flow of water into another person's source of water, eg a well.
Dewatering	The removal of water from a working to keep it dry or to stabilise an embankment.
Flashier	When rain falls on a surface from which vegetation has been removed, it runs off more immediately and more quickly into neighbouring streams instead of providing a steadier supply of water spread over a longer period.
Focal Zones	Variations in weather, temperature and wind speed can affect the way that noise is carried; sometimes noise is focused at places well away from the source, it can be louder than much closer to the noise source – this is a focal zone.
Good Practice	This is a suggestion for a way to ameliorate the effect of an activity on the environment; there is no implication that it will be effective in all situations (see Section 1.4 for a fuller explanation).
Overburden	Soil and other material which overlies the mineral being worked.
Planning Gain	An incidental or planned improvement not directly related to the working.
Plaster Blasting	See Secondary Blasting.
Progressive Restoration	Restoring the site as work proceeds rather than waiting until operations are finished.
Restoration Blasting	A series of blasts designed to improve the post-operational visual appearance and aid revegetation of a quarry face.

Secondary Blasting	The initial blasting may leave lumps of rock which are too large to be handled; explosive can then be attached to these lumps to make them smaller, unfortunately it can be rather noisy and uncertain operation.
Section 106	The Town and Country Planning Act 1990 in Section 106 (as substituted by the Planning and Compensation Act 1991) provides for a developer to enter into an obligation to do certain things related to the development or the land in question, by agreement or otherwise, where the objective cannot satisfactorily be achieved by a planning condition.
Surface Detonating Cord	Explosives can be detonated electrically; sometimes a cord is used which is in itself explosive which burns so quickly that unless it is covered it makes a very loud sharp noise.

17.3 TECHNICAL UNITS

dB	A measure of fluctuating pressure in the air
dB(A)	A measure of fluctuating air pressure corresponding to what people hear
Hz	Frequency of a vibration or noise, cf pitch of a musical note in terms of the number of oscillations per second
L_{90}	The noise level exceeded for 90% of the time
$L_{eq,T}$	A noise level averaged in terms of energy over a period of time 'T'
ms	A thousandth of a second
pH	The measure of acidity/alkalinity
ppv	Peak Particle Velocity is the measure of ground vibration from blasting
μg	A millionth of a gram

CHAPTER 18

REFERENCES (Numerical)

1. County of South Glamorgan, 'County Structure Plan Studies, Local Quarry Plan No 2, Wenvoe/St Andrews Area', Survey Report Feb 1979 & Draft Local Plan Feb 1979 (amended May 1979, Oct 1982 & April 1985)

2. County of South Glamorgan/Mid Glamorgan CC (1982), 'Local Quarry Plan No4. NW Cardiff Radyr/Tongwynlais Area', Joint Consultative Report June 1982

3. Hughes TEV (Inspector), 'Report on the Objections to the Plan made at the Public Local Inquiry [into Radyr/Morganstown District Plan]', Date of Inquiry October 1981, 1982

4. Etherington JR, 'The Effect of Limestone Dust on a Limestone Heath in South Wales', Nature in Wales 15, 1977, pp218-223

5. Leuschner H-J, 'The Opencast Mine of Hambach – A synthesis of raw material winning and landscaping', IXth World Mining Congress II-1, 1986, pp1-17

6. Contractors' Aggregates Ltd, 'Environmental Statement – Land at Chantry Farm/Toppinghoe Hall, Boreham, Essex', published by author, February 1989, pp32

7. UNEP/IEO, 'Environmental Aspects of Selected Nonferrous Metals Ore Mining', Technical Guide 1st draft, October 1989, Paris

8. WS Atkins Engineering Sciences, 'Draft Notes on Planning Guidance', Second Draft, February 1989 (unpublished)

9. Shenton V et al, 'The Control of Substances Hazardous to Health – The experience of a major quarrying group', Quarry Management October 1989, pp25-30

10. Confederation of British Industry, 'Archaeological investigations – Code of Practice for Minerals Operators', CBI Publication March 1991

11. Muskett CJ, 'Environmental Assessment of Opencast Coal Development', Mine and Quarry Environment 2 3, 1988, pp12-15

12. Jay Mineral Services Ltd, 'Mineral Planning Authority Blasting Vibrations Survey', Report by JMS, Bissoe, Truro, Cornwall TR4 8QZ, July 1988

13. Richardson S, 'The Foxhouses Kestrels', Newsletter, Cambria Trust for Nature Conservation, Ambleside, August 88, p3.

14. Cumbria CC, 'Cumbria Coal Local Plan – Draft Written Statement for Consultation', September 1988

15. Cumbria CC & Lake District Special Planning Board, 'Cumbria & Lake District – Joint Minerals Local Plan (Excluding Coal)', Written Statement March 1986

16. Walters SG, 'Buckinghamshire CC Minerals Subject Plan – report of a Public Local Inquiry', Report of Inspector 27th August 1982

17. County Planning Officer, Buckinghamshire CC, 'Report of the County Planning Officer', Planning Sub-committee 13th November 1989

18. Avon, Gloucestershire, and Somerset Environmental Monitoring Committee, 'Vibration. A guide for Environmental Health Officers', Publication available from Hon Secretary, Environmental Health Department, Wansdyke District Council, Midsomer Norton, June 1985

19. Down CG et al, 'The Environmental Impact of Large Stone Quarries and Open-Pit Non-Ferrous Metal Mines in Britain', Final report of the Mining Environmental Research Unit RSM, Imperial College June 1976

20. Anon, 'Woolhampton Quarry – An example of rapid and environmentally sensitive development', M & Q Environment 2 5, 1988, pp26-28

21. Brook C et al, 'Opencast Blasting and the Environment', The Mining Engineer, December 1988, pp253-8

22. Carden S, 'Opencast Mining – Complaints and Nuisance', M & Q Environment 3 2, 1989 p15

23. Walker A, 'Opencast Mining Noise', M & Q Environment 3 2, 1989, pp20-2

24. Woodside A et al, 'Control of Dust in Stone Processing – A survey of Northern Ireland plants', Quarry Management March 1989, pp29-31

25. Broadhurst KA, 'Blast Vibration Evaluation', Paper published by Rock Environmental Limited, Derby, 1987

26. Department of the Environment/Welsh Office, 'Minerals Planning Guidance: Applications, Permissions and Conditions', MPG2, HMSO 1988

27. Department of the Environment/Welsh Office, 'Minerals Planning Guidance: Opencast coal mining', MPG3, HMSO, May 1988

28. Denby B, 'Computer Aids in Quarry Design and Planning – Opencast techniques applied to quarrying', Quarry Management May 1988, pp27-41

29. Jones B, 'Open-cast mining blamed for levels of illness – Doctors say asthma cases 15 times average in town near coal workings', The Times, Monday June 6th 1988, p6

18. References (Numerical)

30. Stephenson IC, 'Are All Quarries Unattractive? Not This One...', *Stone Review*, December 1987, pp11-12

31. Johnson D, 'New Tilcon Quarry Meets Environmental Standards', *Stone Review*, February 1988, pp15-17

32. Holmes PJ, 'The Environmental Impact of Quarrying - Planning procedures and mitigation measures in Australia', *Quarry Management*, April 1988, pp37-43

33. Kestner MO, 'Ten guidelines for developing a plant's dust control system', *Pit & Quarry* c1988

34. County Planning Officer, Hampshire CC, Various Reports by County Planning Officer to the Roads and Development Sub-Committee of Hampshire CC re applications for planning permission for minerals operations 1988-9

35. WS Atkins, 'Noise from Surface Mineral Workings - Community Response Studies: Initial Findings', Draft of work in progress, report to DOE 8/1987 (unpublished)

36. Anon, 'Stateside Stripping - Taywood Mining's environmental experience in America', *M & Q Environment* 1 2, 1987, pp22-24

37. Andrews J, 'When You're Ready to Get Involved with About Face, What Do You Do?', *Stone Review*, June 1987, pp8-9

38. Collens G, 'Coolscar 2: Landscape Issues', *Mineral Planning* 32 September 1987, pp26-9

39. Lambert AK, 'Dust Control at Roadstone Plants', *M & Q Environment* 1 2, 1987, pp14-15

40. Anon, 'Environmental Management at English China Clays', *M & Q Environment*, 1 2, 1987, pp16-18

41. Jarvis D, 'The Use of Computers in Quarry Planning and Design', *M & Q Environment* 1 1, 1987, pp 10-11

42. Corker SP, 'Bulk Soil Handling Techniques', *M & Q Environment*, 1 1, 1987, pp15-16

43. Atlee F, 'Wake Stone: Exemplifying the Good Neighbour Policy and Reaping the Rewards ', *Stone Review*, April 1987, pp10-13

44. Anon, 'The About Face Program: More Than Just Beautification', *Stone Review* April 1987, pp8-19

45. Anon, 'Dust disposal at Halton East Quarry', *Quarry Management* 1987, UK

46. Jenkins B, 'Water and Waste Management for the Pine Creek Gold Mine', *Mine & Quarry* January/February 1987, pp61-62

47. Tuck G, 'The control of Airborne Dust in Quarries', *Quarry Management* April 1987, pp 27

48. Parrett F, 'The Application of Water-Based Foam in Quarry Dust Suppression', *Quarry Management* April 1987, pp39-40

49. Wilson TJ et al, 'Good Vibrations', *Mine & Quarry*, Jan/Feb 1987, pp51-2

50. Anon, 'New Tarmac Quarry in Somerset', *Quarry Management*, March 1986, pp13-18

51. Groundwork Associates Ltd, 'The Minerals Industry Environmental Performance Study' Groundwork Associates Ltd, 1991

52. Atkinson T et al, 'Surface Mining - Future Concepts', Proceedings of symposium 'Surface Mining -Future Concepts' University of Nottingham, 18-20th April 1989, pp5-11

53. Department of the Environment/Welsh Office, 'Mineral Planning Guidance: The review of Mineral Working Sites' MPG 4 HMSO September 1988.

54. Simons DB et al, 'Surface Coal Mining Hydrology', *Minerals and the Environment* 6 4, December 1984, pp133-144

55. Adkins N, 'Aids to Blast Design' *Quarry Management* October 1986, pp35-38

56. Fish B, 'Quarrying and the Community' *Stone Review*, June 1987 pp6-7

57. Stacks JF, 'Stripping - the surface mining of America', a Sierra Club Battlebook, New York, 1972.

58. Council for Environmental Conservation, 'Scar on the Landscape?', Report on Opencast Mining published by CoEnCo London, 1979.

59. Confederation of British Industry, 'Archaeological Remains' Chairman's letter to Members CBI Minerals Committee, 7th April 1982

60. Trettle C, 'Propagation Law and Small Bore Hole Blasting', *Stone Review* June 1986 pp18-19

61. Parkin RJ, 'The Planning, Operational and Environmental Aspects of Opencut Mining at Boundary Hill Mine Central Queensland', *The Mining Engineer* December 1988, pp268-275

62. Laage LW et al, 'Discriminating Back-up Alarm Technology for Front End Loaders', *The Mining Engineer* December 1988, pp276-277

63. Council of the European Communities, 'on The Assessment of the Effects of Certain Public and Private Projects on the Environment' Council Directive 85/337/EEC, 27 June 1985.

64. British Materials Handling Board, 'Code of Practice for the Purchase and Operation of Fabric Filters for Dust Control', British Materials Handling Board August 1985.

65. Tomlinson RV, 'The Public Impact of Quarrying,' *Quarry Management* February 1983, pp85-90

66. Brent-Jones E, 'Opencast Coal Mining in the UK', RICS Conference 22nd October 1981, pp21

67. Norton P, 'Future Trends in Mine Design and Restoration in relation to Environmental Constraints', Proceedings of symposium 'Surface Mining -Future Concepts' University of Nottingham 18-20 April 1989, pp124-127

68. Johnson FM, 'Aspects of Planning in Opencast Mining', From 'Surface Mining - Future Concepts' Proceedings of symposium University of Nottingham 18-20 April 89, pp30-34

69. Department of the Environment/Welsh Office, 'The Reclamation of Mineral Workings' MPG7 HMSO 1989

70. Philipson G, 'Where do we go from here?', *Quarry Management*, May 1988, pp43-46

71. BCOE, 'Complaints in 1989' Internal Memorandum 12 March 1990

72. Purves JB, 'Aspects of mining and pollution control in SW England', IMM Symposium, Minerals and the Environment, London, June 1974, pp159-179

73. Davies IV et al, 'Opencast Coal Mining: working, restoration and reclamation' IMM Symposium, Minerals and the Environment, London, June 1974, pp313-331

74. Moyes AJ et al, 'Economic comparison of dust collection methods in the quarrying industry', IMM Symposium, Minerals and the Environment Symposium, London, June 1974, pp579-588

75. BCOE, 'Complaints in 1990' Internal Memorandum 26 February 1991

76. Platt JW, 'Environmental control at Avoca Mines Ltd. Ireland' IMM Symposium, Minerals and the Environment, London June 1974, pp731-758

77. Paulson KR, 'The Merger of Engineering and Environmental Technology in Modern Mining and Mineral Processing' Industry and the Environment 8 1, 1985, pp2-4

78. Anon, 'Demand Forecasts for Aggregates: "Nightmare Scenario" or Symbol of Prosperity?' ENDS Report 196 May 1991 pp 14-18

79. Commission on Energy and the Environment, 'Coal and the Environment' HMSO 1981, p257

80. Chadwick MJ et al, Environmental Impacts of Coal Mining and Utilisation Land Disturbance and Reclamation after Mining, Pergamon Press 1987 Ch. 2

81. Chadwick MJ et al, Environmental Impacts of Coal Mining and Utilisation Environmental Impacts of Coal Transportation Pergamon Press 1987 Ch. 4

82. Cotton J, 'Archaeological Investigations at Holloway Lane 1980-88', pamphlet pub by Museum of London & Drinkwater Sabey 1988

83. Jenkins B, 'Integrating environmental factors into mining projects' Australian IMM Symposium, Mining and Environment - A Professional Approach, Brisbane July 1987 pp101-108

84. Badham IF, 'Mining and Environmental Noise - Preventing Problems Professionally', Australian IMM Symposium, Mining and Environment - A Professional Approach, Brisbane July 1987, pp109-115

85. Just CD et al, ' The economic and operational implications of blast vibration limits' Australian IMM Symposium, Mining and Environment - A professional Approach, Brisbane July 1987, pp177-124

86. Fish BG, 'Environmental and amenity issues: significance and priorities' Quarry Management August 1983, pp487-492

87. Anon (ENDS), 'The environmental impact of mining and quarrying', ENDS Report 21 March 1979, pp13-15

88. Hibbard WR, 'Environmental Impact of Mining', A Sourcebook on the Environment - a guide to the literature, University of Chicago Press 1987, Ch 14 pp295-308

89. Ketteridge D, 'The Politics of Dust' Mine and Quarry Jan/Feb 1989, pp57-8

90. Dunn P 'Blasting and Vibration control in British Opencast Mining' Mine and Quarry Jan/Feb 1989 pp59-63

91. Bowhill A, 'Balancing Nuisance with Need' Land and Minerals Surveyor 7 1989, pp535-537

92. Cunningham AE, 'Coal Production and the Environment' Department of Energy Library Bibliography 1979

93. Anon, 'European Council of Ministers, Directive on the protection of Workers from Noise' Noise and Vibration Control Worldwide June 1986, p164

94. Local Authority Associations, 'Opencast Mining - Code of Practice' LAA Code, June 1986

95. BC Opencast Executive, 'Opencast Coal Mining in Great Britain' BC Booklet 1988, pp 22

96. Coal Information and Consultancy Services, 'The British Opencast Industry' Report CICS London, March 1989

97. Johnson D, 'New Tilcon Quarry Meets Environmental Standards', Stone Review February 1988 pp15-17

98. Mohanty B et al, 'Optimum Blast Design', The Northern Miner Magazine February 1988, pp23-26

99. Council of European Communities, 'Proposal for a Council Directive on the Conservation of Natural and Semi-Natural Habitats and of Wild Fauna and Flora' Restricted Document 5949/91, Brussels 3 May 1991

100. UK Government 'Town and Country Planning (Assessment of Environmental Effects) Regulations 1988', HMSO 1988

101. Brook C et al, 'Environmental Aspects of Opencast Mining with reference to the Rother Valley Country Park', The Mining Engineer November 1989, pp191-196

102. Thomas W, 'The Landscape Architect and the Coalmining Industry' The Mining Engineer February 1989, pp377-381

103. Froedge DT, 'Outsmarting Blast Vibrations' Coal November 1989, pp67-69

104. Miller GG, 'Selective Overburden Placement' Surface Coal Mining Technology, Engineering and Environmental Aspects, ed Fung R, Noyes Data Corporation, New York, 1981, pp343-347

105. Hill RD, 'The Impacts of Coal Mining on Surface Water and Control Measures Therefor' Surface Coal Mining Technology, Engineering and Environmental Aspects, ed Fung R, Noyes Data Corporation, New York, 1981, pp362-367

106. Martin JF, 'Coal Refuse Disposal Practices and Challenges in the United States' Surface Coal Mining Technology Engineering and Environmental Aspects, ed Fung R, Noyes Data Corporation, New York, 1981, pp368-374

107. Thompson J, 'The Compaction of Opencast Backfill, A new option in restoration strategy', Quarry Management, January 1990, pp11-16

108. Talbot D, 'Recent Advances in Open-pit Blasting, Surface mining experience benefits quarrying', Quarry Management, August 1988, pp31-34

109. McGowan G, 'Coping with Environmental Pressure, The Value of Pre-development Planning and Nuisance Control', Quarry Management, August 1989, pp33-34

110. Bate RR et al, 'Campaigners' Guide to Opencast Mining' CPRE February 1991

111. Kaas LM, 'Bureau of Mines Involvement in Environmental Issues' American Mining Congress, Mining Technology and Policy Issues 1986, Las Vegas October 1986

112. Anon, 'Solving Environmental Problems' World Mining Equipment 9 6, June 1985, pp13-16

113. Kestner MO, 'Enclosure, wet suppression will help control dust', Pit & Quarry January 1988, pp72-74

114. Wiles RA, 'New Publications for Road Systems at Coal Mines' Proceedings 1983 Symposium on Surface Mining, Hydrology, Sedimentology and Reclamation, Kentucky, November 1983, pp283-287

18. References (Numerical)

115. Robinson MK et al, 'Disposal of Excess Spoil: Durable Rockfill' Proceedings 1983 Symposium on Surface Mining, Hydrology, Sedimentology and Reclamation, Kentucky, November 1983, pp179–188

116. van Boven J et al, 'Environmental Compliance at Hoboken in 1983' *Journal of Metals* August 1984, pp62–64

117. Timson JP, 'Environmental Aspects of Mine Planning and Development' Annual Review, Irish Association for Economic Geology, 1988, pp56–60

118. Glover HG, 'Mine Water Pollution – an Overview of Problems and Control Strategies in the United Kingdom' *Water Science and Technology* **15** 2, 1983, pp59–70

119. Zaburunov SA, 'Blasting, Mind over Materials' *Eng & Mining J* April 1990, pp20–25

120. ICI Explosives 'Blast Vibrations Frequency Amplitude Modelling' *Down Line* 12, ICI Explosives, February 1990, pp20–22

121. Burnham DP et al, 'Developments in Blasting Practice' *Quarry Management & Products* October 1981

122. Department of the Environment/Welsh Office, 'Environmental Assessment – A Guide to the Procedures' HMSO 1989

123. Department of the Environment, 'Planning Policy Guidance: Archaeology and Planning' PPG16, HMSO November 1990

124. Williams J, 'Fresh problems springing up' *Quarrying, Construction News Supplement* June 6 1991

125. Hurrell M, 'Extracting industry needs to keeps its footing under exacting environmental demands' *Quarrying, Construction News Supplement*, 6 June 1991 pp4&6

126. Anon, 'Crushing reply to problem of dump trucks' *Quarrying, Construction News Supplement*, 6 June 1991

127. Anon, 'Restoration Bonds put under the Microscope' *Planning*, 19 April 1991

128. Walker A & Cockcroft P, 'Noise Standards –Where are they?' *Mineral Planning* 46 March 1991, pp3–5

129. Anon, 'Commission Changes Tack on Environmental Auditing' *ENDS Report* **194** March 1991 p33

130. Griffiths G, 'Past Perfect? The DoE Planning Policy Guidance on Archaeology and Planning' *Mineral Planning* **46** March 1991 pp9–12

131. Clement K, 'Environmental Auditing for Industry: A European Perspective' *European Environment*, **1** 3 1991, pp1–4

132. Hallett S et al, 'UK Water Pollution Control: A Review of Legislation and Practice' *European Environment* **1** 3 1991, pp7–13

133. Noise Review Working Party, Department of the Environment 'Report of Noise Review Working Party 1990' HMSO 1990

134. Hawkins MR, 'The review of Mineral Working Sites; Results of Questionnaire Survey – Priorities for Action' Report of the County Planning Engineer and Planning Officer to Planning Sub-Committee, Devon CC, EP/91/5/HQ 2 Jan 1991

135. Somerset CC, 'Environment Committee Proposals for Lorries and the Environment' Somerset CC 1991

136. Her Majesty's Inspectorate of Pollution, 'Best practicable means: general principles and practices' HMIP, BPM 1/88, January 1988

137. Department of the Environment/Welsh Office, 'Integrated Pollution Control' A Consultation Paper July 1988

138. Royal Commission on Environmental Pollution, 'Best Practicable Environmental Option' HMSO February 1988

139. International Chamber of Commerce, 'Environmental Auditing' ICC position paper, Paris 1988

140. British Standards Institution, 'Environmental Management Systems' Draft British Standard, EPC/50, Documents 91/53255-7, BSI 1991

141. Commission of the European Communities, 'Draft Proposal for a Council Regulation Establishing a Community Environmental [Management] Auditing Scheme (Eco-Audit)' Document No. XI/83/91 – Rev 2, CEE Brussels May 1991

142. Council of the European Communities, 'Council Directive on the protection of groundwater against pollution caused by certain dangerous substances' Official J of the European Communities No. L 20/43–47, 26 January 1980

143. Department of the Environment/Welsh Office, 'EC Directive on the Protection of Groundwater Against Pollution Caused by Ceertain Dangerous Substances (80/68/EEC)' Circular DOE 4/82, WO 7/82, HMSO March 1982

144. Department of the Environment/Welsh Office, 'EC Directive on the Protection of Groundwater Against Pollution Caused by Certain Dangerous Substances (80/68/EEC): Classification of Listed Substances' Circular DOE 20/90, WO 34/90, HMSO March 1982

145. Tromans S et al, 'The Environmental Protection Act 1990: Its Relevance to Planning Controls' *J of Planning & Environmental Law* June 1991 pp507–515

146. UK Government, 'This Common Inheritance. Britains Environmental Strategy' White Paper, HMSO September 1990

147. CIRIA, 'Construction Industry Research – Rock-Blasting: A Guide to Good Practice' Document 1174/WP1/1, CIRIA September 1989

148. Department of the Environment/Welsh Office, 'Public Rights of Way' Circular DOE/WO 1/83, HMSO February 1983

149. Anon, 'England's garden is handled with care' *Construction Weekly*, 12 June 1991, pp18–19

150. Secretary of State for the Environment decision on, 'Town and Country Planning Act 1971: Section 36. Land at Bridge Farm, Willingham Appeal by Tarmac Roadstone Limited' APP/W0530/A-/88/83638, 15 March 1989

151. Ripley EA et al, *Environmental Impact of Mining in Canada*, Centre for Resource Studies Queens University Ontario, 1978

152. White T, 'Blasting to Specification' *Quarry Management* December 1989, pp27–34

153. Rendel S et al, 'Planning Consent for Mineral Extraction and Landfill: The Wildlife Factor' LEB winter 1990/91, pp6–8

154. Smith M, 'North Pennines Landscape Policy' *Mineral Planning* December 1989, pp25–26

155. Murthy RK et al, 'South Arcot Aquifer and its Utilisation' *Braunkohle* **38** 8, August 1986, pp223–226

156. OECD, Environment Committee, Group on the Energy, 'Coal and Environmental Protection (Costs and Costing methods)' report Paris, 1983

18. References (Numerical)

157. OECD, Environment Committee, Group on the Energy, 'Coal and Environmental Protection (Environmental issues and remedies)', report Paris, 1983

158. The Opencast Mining Intelligence Group, 'A Reassessment of Opencast Mining' Report for a conference on opencast mining in 1979

159. Tollner EW et al, 'The use of grass filters for sediment control in strip mine drainage Volume 1', Report prepared by University of Kentucky Institute of Mining and Mineral Research, 1987

160. UN ESCAP, 'Proceedings of the working group meeting on environmental management in mineral resource development' Mineral Resource Development Series, 49, 1982

161. Department of the Environment et al, 'Coal and the Environment', Government's response to the Commission on Energy and the Environment's Report, HMSO London, 1983

162. Anthony Goss Associates et al, 'Planning Conditions for Hard Rock Quarries' Report for the DoE 1983

163. Royal Society for Nature Conservation, 'Losing Ground – Skeletons in the Cupboard – Mineral Planning' RSNC March 1991

164. Coalfield Communities Campaign, 'Opencast Coal', Report by the CCC Opencast Coal Working Party, 1987

165. Geoffrey Walton, 'Review of current practice in British quarries etc.' Tasks 2&3 of DoE commissioned report, February 1986

166. Geoffrey Walton, 'Handbook on the hydrology and stability of excavated slopes in quarries', Task 5 of DoE commissioned report, 2nd impression 1988

167. Geoffrey Walton, 'The stability and hydrogeology of mineral workings', Summary of findings of previous tasks commissioned by Department of the Environment, July 1986

168. IEA Coal Research, 'Coal Research Projects 1988', Abstracts of current and recent technology R&D 1988, pp440-458

169. Anon, 'Why harm nature for peat's sake?' *Independent*, 11th December 1989

170. McCarthy M, 'Quarry firms warned over damage to countryside' *The Times* 3 May 1991

171. Pollock SHA, 'Coastal Superquarries – from concept to reality. Planning implications of the Glensanda development, Highland Region, Scotland', IMM Symposium, Minescape '88, 1988, pp313-321

172. Riddleston B, 'Opencast Coal Mining in South Wales – Environmental Issues in Practice' IMM Symposium, Minescape '88, 1988, pp179-189

173. Bate RR, 'Opencast Coalmining: The view of the CPRE' IMM Symposium, Minescape '88, 1988, pp171-177

174. Beeley JG, 'Instrumentation and the Surface Environment', IMM Symposium, Minescape '88, 1988, pp53-67

175. Miles L, 'Waging Coal War', *Surveyor* 25th February 1988 pp12-15

176. Down CG et al, *'Environmental Impact of Mining'* Ch 2 'Range and Importance of Environmental Problems', Applied Science Publishers 1977, pp10-27

177. Down CG et al, *'Environmental Impact of Mining'* Ch 3 'Visual Impact', Applied Science Publishers 1977, pp28-56

178. Down CG et al, *'Environmental Impact of Mining'* Ch 4 'Air Pollution', Applied Science Publishers 1977, pp57-88

179. Down CG et al, *'Environmental Impact of Mining'* Ch 5 'Water Pollution', Applied Science Publishers 1977, pp89-144

180. Down CG et al, *'Environmental Impact of Mining'* Ch 6 'Noise' Applied Science Publishers 1977, pp142-163

181. Upton SL et al, 'Some Experiments on Material Dustiness', Paper for presentation at the Aerosol Society Annual Conference Univ Surrey April 1990

182. Nalpanis P et al, 'Suspension, Transport and deposition of Dust from Stockpiles', Report from Warren Spring for DTI, November 1988, p71

183. NCB Mining Department, 'Water in the Coal Mining Industry (Technical Management of)' NCB 1982, Ch 18

184. Lane DD et al, 'Use of Laboratory Methods to Quantify Dust Suppressant Effectiveness', *Trans Society of Mining Eng of AIME* **274**, pp2001-2004

185. Libicki J, 'Changes in Groundwater due to Surface Mining', *Int Journal of Mine Water*, **1** 1982, pp25-30

186. Lindorff DE, 'Hydrogeology of Surface Mine Spoil in Illinois' Annual Meeting Society of Mining Engineers of AIME, Illinois, Preprint 81-24 February 1981

187. Anon, 'Tidying up the Mess of Britain's Discharge Consent System' *ENDS Report* **186** July 1990 pp17-24

188. Smith M, 'Taking nature for a ride' *Guardian*, 31 August 1990

189. Cowe R, 'Greening by numbers: a new role for accountants' *Guardian* 19 June 1990

190. Cross D, 'A Slippery Situation' *Surveyor* **1** March 1990

191. North R, 'Quarry companies present new face' *Independent* 23 January 1990

192. Colley R et al, 'Reviewing the Quality of Environmental Statements' *The Planner* 27 April 1990, pp12-13

193. Anon, 'World Lead Claimed for British Standard on Environmental Management' *ENDS Report* **197** June 1991 p4

194. Commission of the European Communities, 'A Consultation Paper on Draft Elements for a Council Directive on the Environmental Auditing of Certain Industrial Activities' EEC, XI/632/90-EN December 1990

195. Anon, 'Emerging Issues in Environmental Auditing' *ENDS Report* **185** June 1990 pp 11-13

196. Anon, 'Brussels takes a First Step towards an Environmental Auditing Directive' *ENDS Report* **192** January 1991, pp13-16

197. Anon, 'First Draft Guidance on Integrated Pollution Control' *ENDS Report* **188** September 1990 pp27-28

198. Anon, 'Quarry Planning: Is There a Role for Computers?' *Mine & Quarry* May 1990 pp19-23

199. Anon, 'National Parks: Review and Minerals Policies' *Mineral Planning* **44** September 1990 pp3-10

200. Anon, 'Mendip Limestone: And then there was one...' *Mineral Planning* **47** June 1991 p17

18. References (Numerical)

201. Secretary of State for the Environment decision on, 'Coolscar Quarry, Kilnsey, North Yorkshire, Eskett Quarries Limited', M/5069/42/4 10th March 1987

202. Williams J, 'Fugitive Dust', *Mine and Quarry*, March 1986, pp9-10

203. West J, 'Involve Employees to Build Support – Public Acceptance', *Rock Products*, March 1986, pp54-57

204. Hodges DJ et al, 'Computer graphics for public enquiries', *Colliery Guardian*, February 1986, pp56-62

205. Rawlinson RD et al, 'Noise from Surface Mineral Workings ', *Mine & Quarry* April 1986, pp26-28

206. Thomas JB, 'Developments in Explosives Utilization in ECC', *Quarry Management*, 1986

207. Anon, 'Success of US About Face Programme', *Quarry Management*, April 1988, pp45-46

208. Anon, 'A Clean Sweep for Blue Circle', *Quarry Management*, April 1987, p40

209. Wilson H, 'A look at current and future trends', *Quarry Management*, April 1988 pp33-35

210. Stacks JF 'Social Impact', *Stripping*, Chapter 4, Sierra Club New York 1972, pp51-66

211. Hampshire CC, "Model" planning conditions', Publication 1/1988, Hampshire CC, January 1988

212. Heasman IN, 'Mega-Quarries – The environmental impact of the trend towards giantism' *M & Q Environment* 3 1, 1989, pp23-27

213. Clark AR et al, 'Magnesian Limestone Quarrying Planning and Environmental Considerations', *M & Q Environment* 3 1, 1989, pp15-17

214. Ellis AF, 'The Control of Air Pollution in the Minerals Industry', Proceedings of 49th NSCA Annual Conference Llandudno 19th October 1982, NSCA Brighton 1982

215. Anon, 'Dust Control', *Mining Magazine*, June 1979, pp504-521

216. Anon, 'Pont Einion and the Environment', *Colliery Guardian* July 1988, pp226-7

217. South Glamorgan County Council, 'Progress to Date 1976-1984', Internal Report South Glamorgan County Council 1984, pp173-204

218. Golder Associates (UK) Ltd, 'Blast Diagnosis Service', Brochure Published by Golder Associates (UK) Ltd February 1989

219. Griffin CG, 'Ground Vibration and Air Overpressure from Blasting', Guidance Note, Cornwall County Council, July 1989

220. Atkins Research and Development, 'Environmental Impact of Mining and Quarrying Activities', Monograph December 1980

221. Kerry G, 'Dealing with Quarry Blast Noise', Proceedings of Symposium 'The Problem of Assessing Environmental Noise', The Association of Consulting Scientists, London, 25th March 1981

222. Broadhurst K et al, 'Review of Current Standards and Recommendations for Vibration and Noise', IMM Conf 14/2/1984

223. Wilton TJ, 'Air Overpressures from Blasting' Paper published by Rock Environmental Ltd, undated

224. Broadhurst KA et al, 'Blast Induced Ground Vibration and Air Overpressure– Terminology, Standards and Control', Paper published by Rock Environmental Ltd, undated

225. Anon, 'Code of Good Practice – Ground and Airborne Vibration From the Use of Explosives', Publication, Federation of European Explosives Manufacturers, Brussels 1985, pp21

226. Anon, 'Report of the Commission on Mining and the Environment' HMSO 1972

227. Finlayson D, 'Ballymoney Lignite Project – Hydrology and Environmental Assessment', *M & Q Environment* 3 2, 1989, pp17-19

228. Peart JD et al, 'Authorisation Procedures for Opencast Mining', *Journal Planning & Environmental Law*, January 1985, pp11-20

229. Jarvis D, 'Visual Impacts – Who Sees What from Where?', *Minerals Planning*, 26 March 1986, pp8-9

230. Offord C et al, 'Open Cast Coal Mining – A Local Authority Viewpoint', *M & Q Environment* 3 2, 1989, pp10-14

231. Anon, 'Go ahead for Cheltenham extraction/landfill site', Planning News, *M & Q Environment* 3 2, 1989, p6

232. County Planning Officer, County of South Glamorgan, 'Quarry Monitoring Programme', Report of County Planning Officer to Environmental Planning Committee, 21st May 1979

233. Attewell PB, 'Noise and vibration in civil engineering', *Municipal Engineer* 3 June 1986, pp139-158

234. Nguyen YV et al, 'Identification and Control of Sources of Airborne Coal Dust at Large Storage Coal Piles', paper 80-11.4 73rd Annual Meeting of the Air Pollution Control Association, Montreal June 22-27, 1980

235. Chandler AJ et al, 'The Application and Verification of Open Source Fugitive Emission Estimates', paper78-55.5, 71st Annual Meeting of the Air Pollution Control Association, Houston June 25-30 1978

236. Hancock RP, 'Visual Response to Dustiness', *Journal of the Air Pollution Control Association*, January 1986 **26** 1, 1986, pp54-57

237. Anon, 'Effects of Particulate Air pollutants on Vegetation', Report of Subcommittee on Airborne Particles, Committee on Medical and Biologic Effects of Environmental Pollutants, Division of Medical Sciences, Assembly of Life Sciences, National Research Council; University Park Press, Baltimore Ch9 Airborne Particles 1979, pp199-214

238. Canessa W, 'Dust Retardants ', Chapter 16 of 'Air Pollution Control and Design Handbook' Part 1 Edited by Cheremisinoff PN et al, Dekker 1977, pp431-447

239. Bennett JH et al, 'Assessment of Fugitive Emissions from Sand and Gravel Processing Operations', 73rd Annual Meeting of the Air Pollution Control Association, Montreal June 22-27, 1980, pp14

240. Allett EJ, 'Bretby and the Environment', *Colliery Guardian* March 1989, pp84-85

241. Anon, 'Ancient Monuments in the Countryside', English Heritage Publication 1987, pp15

242. Secretary of State for the Environment decision on, 'Appeal by Redland Aggregates Ltd. Application No.N86/577', 12th December 1988

18. References (Numerical)

243. Miller ML et al, 'Mendham Priory, Suffolk, proposed extraction of sand and gravel at Priory Farm', Correspondence 15th September 1989, 5th January 1989 and 7th November 1988 supplied by English Heritage

244. Department of the Environment, MAFF and Welsh Office, 'The Water Act – Code of Practice on Conservation, Access and Reacreation' DOE, July 1989, pp12–16

245. Park DG, 'Relocating magnesian limestone grassland' *Biological habitat reconstruction*, ed Buckley GP, Belhaven Press London 1989

246. Department of the Environment/MAFF, 'Conservation Guidelines for Water Authorities, Internal Drainage Boards, Local Authorities, The Nature Conservancy Council and the Countryside Commission', Chapter 7 Protection of Archaeological and Other Historic Features, MAFF/DoE Publication 1988, pp15–16

247. Down CG, 'The Design and Construction of Amenity Banks and Tips – with particular reference to landscaping', Report to Department of the Environment, Royal School of Mines, September 1976, pp243

248. Fleming BF, 'A Study of the Effects of Quarries on Adjacent Woodland', Report for MSc, Centre for Environmental Technology, Imperial College, London, 1986, pp124

249. Grigg CFJ, 'Landscaping Techniques and Restoration', *Mining Magazine*, December 1988, pp492–497

250. Anon, 'Ffos Las opencast coal mine, South Wales', *Mining Magazine*, January 1987, p8

251. Grimshaw PN, 'Butterwell Mine, UK has much to Celebrate', *Mining Magazine*, January 1987, pp28–37

252. Dumenil P, 'The use of explosives in quarries' Blue Circle notes for Meeting Institute of Quarrying, 9th November 1989

253. Kerry G et al, 'Towards a Greater Understanding of Sound Levels from Blasting', *Quarry Management and Products*, April 1983, pp213–217

254. Oates JAH et al, 'The Environmental Effects of Sound from Quarry Blasting', unknown origin

255. Dalgleish I, 'Cost-Effective Drilling and Blasting – Developments contribute to efficiency and safety', *Quarry Management*, January 1989, pp19–27

256. Rukavina M, 'Sand and Gravel Scrubbed/Upgraded Without Water', *Rock Products*, October 1989, p21

257. Stagg MS et al, 'Effects of Repeated Blasting on a Wood-Frame House', US Bureau of Mines Report of Investigations, RI8896, 1984

258. Ball M, 'A Review of Blast Design Considerations in Quarrying and Opencast Mining – Part 1 Principles of design, Part 2 Initiation and computerized design', *Quarry Management* June 1988, pp35–39 and July 1988, pp23–27

259. GLC, 'Beddington Farm Lands – Draft Planning Brief for Gravel Extraction and Restoration', GLC, BMA05, 1983

260. Rock GA, 'The Environmental Noise Impact of a Large Quarry Extension', Proceedings IOA Meeting 'Noise from drilling, mining and quarrying operations', Bournemouth, April 1989

261. Rock GA, 'Planning and Noise from Mineral Workings', Proceedings IOA 11 pt 6, 1989, pp27–34

262. Wells HD, 'An Evaluation of the Sources and Effects of, and Control Mechanisms for, Particulate Emissions ....re Delabole Slate Quarry Cornwall', MSc thesis, Centre for Environmental Technology, Imperial College, London, September 1985

263. Preston CJ, 'Use of Chemical Binders in Dust Control', MSc thesis, Centre for Environmental Technology, Imperial College, 1980

264. Fuchs P, 'To quarry or not to quarry... Blot on landscape', *Construction News*, January 11th 1990, pp16–19

265. Anon, 'No 'Mappa Minerals' in Lugg Valley', Recent mineral cases, *Mineral Planning* 40, September 1989, pp27–28

266. Anon, 'Who is Saying What About Mineral Planning', Bodytalk, *Minerals Planning* 40, September 1989, pp35–36

267. WS Atkins Engineering Sciences Ltd, 'The Control of Noise at Surface Mineral Workings', Report on behalf of Department of the Environment, HMSO London 1990

268. The Trust for Wessex Archaeology, 'Lower Farm, Greenham Archaeological Evaluation', Report 1987 published by The Trust for Wessex Archaeology

269. British Standards Institution, 'British Standard Guide to 'Evaluation of human exposure to vibration in buildings (1 Hz to 80 Hz)'', BS 6472:1984 BSI London, 1984

270. New BM, 'Ground Vibration Caused by Civil Engineering Works', Research Report 53 of Transport and Road Research Laboratory, Crowthorne, 1986

271. Fletcher LR et al, 'Reducing accidents through improved blasting safety', Part of Information Circular 9135, 'Surface Mine Blasting', Proceedings: Bureau of Mines Technology Transfer Seminar, Chicago, 1987, pp6–14+

272. Peltier MA et al, 'Blaster's Training Manual for Metal and Nonmetal Miners', Part of Information Circular 9135, 'Surface Mine Blasting', Proceedings: Bureau of Mines Technology Transfer Seminar, Chicago, 1987, p19–24

273. Stagg MS et al, 'Effects of Blast Vibration on Construction Material Cracking in Residential Structures', Information Circular 9135, 'Surface Mine Blasting', Proceedings: Bureau of Mines Technology Transfer Seminar, Chicago, 1987, pp32–43

274. Kopp JW, 'Initiation Timing Influence on Ground Vibration and Airblast', Information Circular 9135, 'Surface Mine Blasting', Proceedings: Bureau of Mines Technology Transfer Seminar, Chicago, 1987, pp51–59

275. Stagg MS et al, 'Influence of Blast Delay Time on Rock Fragmentation: One-Tenth-Scale Tests', Information Circular 9135, 'Surface Mine Blasting', Proceedings: Bureau of Mines Technology Transfer Seminar, Chicago, 1987, pp79–95

276. Kopp JW, 'Stemming Ejection and Burden Movements of Small Borehole Blasts', Information Circular 9135, 'Surface Mine Blasting', Proceedings: Bureau of Mines Technology Transfer Seminar, Chicago, 1987, pp 106–114

277. MJ Carter Associates, 'Low Level Restoration of Sand and Gravel Workings to Agriculture with Permanent Pumping', Report to Department of the Environment, HMSO October 1988

278. Redman M, 'Archaeology and Development', *Journal of Planning and Environmental Law* January 1990, pp87–98

279. Anon, 'Britain – DoE's Near-final List of Industries to be Controlled by Green Bill', *ENDS Report* 180 January 1990, p27

18. References (Numerical)

280. Somerville SH, 'Control of Groundwater for Temporary Works', Report 113, CIRIA, London, 1986

281. Hantz D et al, 'Principaux effets de l'exploitation des carrières sur l'environnement immédiat', *Mines et Carrières* **71**, May 1989, pp46–50

282. Waller RA, 'A Guide to reducing the exposure of construction workers to noise' CIRIA report 120, London, 1990

283. Anon, 'The use of screens to reduce noise from sites', CIRIA Practice Note, Special Publication 38, London, 1985

284. British Standards Institution, 'Noise control on construction and open sites', British Standard BS5228: Parts 1–3: BSI London 1984

285. Blue Circle Industries plc, 'Hope', Proposal produced by Blue Circle Industries 1988 relating to Hope Cement Works 1988

286. Anon, 'Time to Speak Out' *Mining Magazine* June 1990

287. Doncaster MBC Planning Directorate, 'Levels of Vibration and Air Blast at Warmsworth Quarry', Report 1990

288. McCarthy RE, 'Surface Mine Siltation Control', *Mining Congress Journal*, June 1973, pp30–35

289. Burnham DP et al, 'Developments in Blasting Practice', *Quarry Management and Products*, February 1982, pp95–102

290. Ancich EJ, 'The Environmental Aspects of Structural Response to Blasting Overpressure', *Quarry Management and Products* July 1983, pp420–425

291. Barnett JL et al, 'Fugitive Emissions – a Directional Sampler for Particulates', Warren Spring Laboratory report published by Department of Trade and Industry, Stevenage, 1987

292. BCOE, 'Complaint Report', BCOE 1990

293. Anon, 'Proposed Opencast Mining at Potatopot', Material for Exhibition, Workington 1982

294. Turland MJ, 'Planning Permission Conditions – are they a Waste of Resources?', Correspondence, *Journal of Planning and Environment Law* March 1990, pp184–5

295. Anon, 'Applications of Remote Sensing to Environmental Aspects of Surface Mining', Surface Mining, EOSAT Landsat Applications Issue 2, NRSC Farnborough, undated

296. Gilbert OL, 'An Alkaline Dust Effect on Epiphytic Lichens', *Lichenologist*, **8** 1976, pp173–178

297. Department of the Environment/Welsh Office, 'Planning and Noise', Circular DOE 10/73, WO 16/73, HMSO 1973

298. Cumbria CC, 'Planning permission for Potatopot', Reference 2/84/0749, 8th June 1985, letter to NCBOE

299. British Standards Institution, 'Method for Rating Industrial Noise Affecting Mixed Residential and Industrial Areas', British Standard 4142: 1990, BSI London 1990

300. British Standards Institution, 'Methods for the Measurement of Air Pollution', British Standard 1747:1972, BSI London 1972

301. Schofield C et al, 'Guide to the handling of Dusty Materials in Ports', British Materials Handling Board, April 1983

302. Anon, 'Brightsite', Campaign Information Pack, Groundwork, undated

303. Civic Trust, County Surveyors Society, Department of Transport, 'Lorries in the community', HMSO, 1990

304. Anon, 'Opencast Failure' *Mineral Planning* **38** March 1989, pp28–29

305. Anon, 'Recent mineral cases', *Mineral Planning* **38** March 1989, pp14–18

306. Ellis DM et al, 'A Candle in the Wind?' *Mineral Planning* **38** March 1989, pp3–13

307. British Standards Institution, 'Methods for the measurement of Air Pollution – Part 5 Directional Dust Gauges' BS1747:1972, BSI London, 1972

308. Ralph M O et al, 'Performance of the British Standard Directional Dust Gauge' Warren Spring Laboratory Paper No W89001(PA), Stevenage, January 1989

309. Health and Safety Executive, 'Explosives at Quarries – Quarries [Explosives] Regulations 1988 – Approved Code of Practice' HMSO, London, 1989

310. Tomlinson P, 'The Environmental Impact of Opencast Coal Mining' Town *Planning Review*, **53** 1, Jan 1982, pp5–28

311. Ball M J et al, 'Field Experience with new Methods of Electric Shotfiring' 8th Annual Conference of the Society of Explosives Engineers, New Orleans, February 1982

312. Brandt CJ et al, 'Effects of Limestone Dust Accumulation on Lateral Growth of Forest Trees' *Environmental Pollution* **4** 1973 pp207–213

313. Brandt CJ et al, 'Effects of Limestone Dust Accumulation on Composition of a Forest Community' *Environmental Pollution* **3** 1972 pp217–225

314. Anon, 'Bigger Private Opencast Coal Mines face New Environmental Duty' *ENDS Report* **182** March 1990 p27

315. Grimshaw PN, 'Environmental Benefits of Surface Mining' *Mining Magazine* December 1986 pp581–5

316. Gowan M, 'Water in the China Clay Industry' *Journal IWEM* **1** 1, August 1987, pp123–128

317. British Standards Institution, 'Guide To Evaluation of Human Exposure to Vibration in Buildings (1Hz to 80Hz)' Draft Standard (Revision of BS 6472 : 1984), BSI London, 3 February 1990

318. Offord C et al, 'Planning and Opencast Coal : Stobswood, Northumberland' *Mineral Planning* **39** June 1989 pp6–9

319. Anon, 'Legislation – Planning Gain and Agreements' *Mineral Planning* **40** September 1989 pp23–24

320. Hall DJ et al, 'An Improved, Readily Available Dustfall Gauge' *Atmospheric Environment* **20** 1 pp219–222

321. Groom AR, 'Planning Conditions and the Quarry Manager' *Quarry Management and Products* July 1982, pp5

322. Health and Safety Executive, 'Fly Rock Projection from Quarry Blasting' Quarries Topic Report, Health and Safety Executive, September 1989

323. Anon, 'Lorry Sheeting' *Mineral Planning* June 1990 pp12–14

324. Hammersley R, 'Opencast Coal: A genuine cause for concern?' *Mineral Planning* March 1990 pp5–8

325. Stanton W, 'Hard Limestone: Too valuable to quarry' *Mineral Planning*, June 1990 pp3–9

326. Gosling D, 'Exposing Some of the Myths about Opencasting – A Case Study' *Mineral Planning*, March 1990 pp8–10

18. References (Numerical)

327. Wilton TJ, 'Marley Hill Opencast Site' Proof of Evidence Public Enquiry, Rock Environmental Ltd, 31st October 1989

328. Forestry Commission, 'Catalogue of Publications' Forestry Commission, Farnham 1991

329. Anon, 'The Siting and Construction of Temporary Spoil Mounds at Opencast Coal Sites' Code of Practice published by the Federation of Civil Engineering Contractors, reissued 1989

330. Anon, 'The Stability of Excavated Slopes at Opencast Coal Sites' Code of Practice published by the Federation of Civil Engineering Contractors, October 1989

331. Anon, 'The Original Smart Alarm Reversing Bleeper' Leaflet/brochure published by Brigade Electronics Ltd, London, undated

332. Anon, 'RWA 480 Electronic Reverse Warning Alarm' Data sheet 004 of FAW Electronics Ltd, Chesterfield, undated

333. Griffin MJ et al, 'Discomfort produced by impulsive whole-body vibration' *Journal of Acoustical Society of America* **68** 5, November 1980, pp1277-1284

334. Wiss J F, 'Construction Vibrations: State-of-the-Art' *Journal of the Geotechnical Engineering Division, Proceedings ASCE* **107** GT2, February 1981, pp167-181

335. Pearce D et al, 'Blueprint for a Green Economy' Report to DoE published by Earthscan Publications Ltd, London, 1989

336. Dowding CH et al, 'Simultaneous Airblast and Ground Motion Response' *Journal of the Structural Division, Proceedings of ASCE* **108** ST11, November 1982, pp2363-78

337. Bristow C, 'How New Patterns of Demand for Minerals are Created by Changing Socio-economic Trends' Science 89, Annual Meeting of British Association, Sheffield September 1989 Paper No Gl.12

338. British Aggregates and Construction Materials Industry, 'Why Quarry? - 6 - The Caring Face of Quarrying' A BACMI Information Sheet, London 1989

339. Phillipson R (Interview), 'Industry planning to beat shortfall' *Construction News*, Special Report on Quarrying and Mineral Extraction, June 28 1990 p23

340. McLoughlin J, 'BACMI winning on aggregate' Construction News, Special Report on Quarrying and Mineral Extraction June 28th 1990 p26

341. Anon, 'An Award Winning Restoration Scheme - Horton Trout Fishery - Haul Aggregates' *Construction News* June 28th 1990, Special Report on Quarrying and Mineral Extraction p26

342. Anon, 'Digging in for growth' *Construction News* June 28th 1990 Special Report Quarrying and Mineral Extraction p28

343. Higgins T, 'SAGA backs green agenda' *Construction News*, June 28th 1990 Special Report, Quarrying and Mineral Extraction p30.

344. Dyson S, 'Quarrying takes flight' *Construction News*, June 28th 1990 Special Report, Quarrying and Mineral Extraction p30.

345. Anon, 'Clay waste mountain more than a molehill' *Construction News*, June 28th 1990, Special Report Quarrying and Mineral Extraction, p32.

346. Anon, 'C-Vision has safety in view' *Construction News*, June 28th 1990 Special Report Quarrying and Mineral Extraction p34.

347. Anon, 'Thames Water makes a splash' *Construction News* June 28th 1990 Special Report Quarrying and Mineral Extraction p34.

348. Department of the Environment/Welsh Office, 'Town and Country Planning Act 1971, Planning Gain' Circular DOE 22/83, WO 46/83, HMSO August 1983.

349. Carnwath R, 'Enforcing Planning Control' Report to the Department of the Environment, HMSO, Feb 1989 .

350. Machin AW, 'Cambridgeshire County Council Appeal by Tarmac Roadstone Ltd. Report of Inspector to Secretary of State' B/314/SM/P 1st December 1988 (Sand & Gravel, Bridge Farm, Willingham)

351. Donnison RD, 'Derbyshire County Council Appeal by British Coal Corporation (Opencast Executive)' Report of Inspector to Secretary of State for Env, ref APP/D1000/A/88/87968, 14th February 1989 (opencast coal and clay, Kirk & Ryefield)

352. Donnison RD, 'Staffordshire County Council, Appeal and Applications by The British Coal Corporation: Opencast Executive' Report to Secretary of State for the Environment on the Brown Lees opencast working, Ref A/139X/KP/P, 30 Nov 1987.

353. Kemmann-Lane T, 'Peak Park Joint Planning Board, Appeal under Section 36 of the Town and Country Planning Act 1971 by T W Ward (Roadstone) Ltd. against the refusal of planning permission for the quarrying of limestone at Eldon Hill Quarry' Report of Inspector to the Secretary of State for the Environment, Reference A/557X/AJT/P, 17th December 1986.

354. Cookson T, 'Dartmoor National Park, Appeal by British Rail Property Board' Report to Secretary of State for the Environment re Meldon Quarry, Okehampton, Ref A/621X/JLC/P 22nd January 1988.

355. Smith BH, 'North Yorkshire CC, Yorkshire Dales National Park, Appeals and Applications by Eskett Quarries Ltd. (Reopened Inquiry)' Report to the Secretary of State for the Environment re Coolscar Quarry, Ref C/783/MB/P 17th August 1985.

356. Walker PD, 'Cheshire County Council Application by Dalefords Estates Ltd.' Report to Secretary of State re Moss Farm, Sandiway, Ref A/257X/KP/P December 1985.

357. Cheer JS, 'Berkshire County Council Appeal by S Grunden (EWELME) Ltd.' Report to Secretary of State for the Environment re extraction of sand at Old Kiln Farm, Chieveley, Newbury, ref DPI101/P6/ST/P October 1988.

358. Kemmann-Lane T, 'Northumberland C C Appeal by British Coal Corporation' Report to the Secretary of State for the Environment regarding Linton Opencast Workings Ref D/1X/HB/P 6th September 1988.

359. Secretary of State for the Environment, decision on 'Appeal by British Coal Corporation' APP/D1000/A/88/87968 July 1989 relating to Kirk Opencast Site

360. Secretary of State for the Environment, decision on 'Appeal by British Coal Corporation' APP/N3400/A/86/59987, 6th April 1988.

361. Secretary of State for the Environment, decision on 'Coolscar Quarry, Kilnsey, North Yorkshire Eskett Quarries Ltd' M/5069/42/4 10th March 1987 related to Coolscar Quarry.

362. Secretary of State for the Environment, decision on 'Appeal by British Coal Corporation APP/R2900/A/88/85681', 21st December 1988 related to Linton Colliery.

18. References (Numerical)

363. Bollinger GA, 'Blast Vibration Analysis' Southern Illinois University Press, Carbondale, 1971.

364. Davies MR et al, 'An Application for Planning Permission to Extend Sand and Gravel Quarrying Operations at Lydd, Kent and Camber, East Sussex – 2 Environmental Statement, Publication by Brett, August 1989.

365. Carpenter SL, 'Environmental Conflicts in Mineral Development or The Art of Creating Unnecessary Problems' *Minerals and the Environment*, **2** 1989 pp159–164.

366. Pearse G, 'Belt conveyors – increasingly the chosen option' *Mining Magazine* April 1984 pp362–373.

367. Anon, 'The Influence of Quarrying on Groundwater Resources' *Quarry Management and Products*, April 1979 pp97–99.

368. Anon, 'Aerial ropeways – An efficient and economic transport system' *Quarry Management and Products* December 1981 pp 844–847.

369. Anon, 'Tarmac Railhead in Kent – Automated control systems at Hothfield' *Quarry Management and Products*, June 1981, pp403–408.

370. Anon, 'How the environment was improved at Markle Mains Quarry' source unkown, March 1981 p9.

371. Roberts DI, 'Construction Aggregates and the Cost to Conservation' *Quarry Management and Products*, March 1981 pp167–174.

372. Ackroyd T, 'High Hopes for Low Profile' *Mineral Planning* **3** June 1980 pp1,8,9

373. Anon, 'Opencast Coal Productivity Maintained Despite Environmental Restrictions' *Quarry & Mining News*, No.96 January 18th 1980, p1

374. Anon, 'New Wimpey Plan to Secure Jobs and Improve Environment' *Quarry Management* July 1990 p37.

375. Anon, 'Steetley Complete Plantation Removal – Rare species preserved in Thristlington project' *Quarry Management* August 1990 p31

376. Anon, 'Burlington Slate Invest in New Production Methods – 'Green' techniques improve yield' *Quarry Management* July 1990 p35.

377. Siskind DE et al, 'Noise and Vibrations in Residential Structures From Quarry Production Blasting' US Bureau of Mines Report of Investigations 8168, Washington 1976

378. Millington GR, 'Environmental Problems Associated with Opencasting' *Clean Air* **11** 2, 1981, pp47–54

379. Jackson JR, 'Environmental Problems Associated with Opencasting – a Selective Review of Practical Considerations' *Clean Air* **11** 2, 1981, pp66–67

380. Garner K, 'Controlling the Quarry Dust Hazard' *Quarry Management and Products* November 1983 p753

381. Shukla J et al, 'Effect of Cement Dust on the Growth and Yield of *Brassica campestris* L.' *Environmental Pollution* **66**, 1990, pp81–88.

382. Finley BJ 'An Application for Planning Permission to Extend Sand Quarrying Operations into Foxes Bank – Ryarsh, Environment Assessment **2**' Published by Ryarsh Brick Ltd, March 1990

383. Tyldesley D et al 'Ling Hall Quarry, Lawford Heath Warwickshire – Environmental Impact Statement' for Ideal Aggregates Ltd, March 1990

384. RPS Group PLC 'Planning Application and Environmental Impact Statement for the Extraction of Sand and Gravel from Land North of Hertford' Written Statement and Appendices, non-technical Summary, RPS Group plc, May 1990

385. Waller RA 'Planning to Avoid Noise Exposure in Construction' Construction Industry Research & Information Association Technical Note TN 138, 1990

386. Hardaway J 'Coal Mining and Ground Water' *Surface Coal Mining Technology – Engineering and Environmental Aspects* ed Fung R, Noyes Data Corporation, NJ USA, 1981, pp 350–361

387. ARC Ltd for Stone Federation, private communication July 1991

388. Beaman AL 'NTPC Ash Disposal – Review of Soil Stabilisers' Report by W S Atkins & Ptnrs September 1988

389. Beaman AL et al 'Recent Developments in the Method of Using Sticky Pads for the Measurement of Particulate Nuisance' *Clean Air* **14** 2, 1984, pp74–81

390. Beaman AL et al 'Assessment of Nuisance from Deposited Particulates Using a Simple and Inexpensive Measuring System' *Clean Air* **11** 2, 1981, pp 77–81

391. Alabaster JS 'River Pollution by China Clay Wastes and Solids in Suspension' *Environmental Management of Mineral Wastes*, ed Goodman GT et al, Sijthoff & Moordhoff, Netherlands 1978 pp237–238

392. Coppin N, et al 'Environmental Assessment for Opencast Coal Mining' *M & Q Environment* **3** 3, 1989, pp 17–22

393. Bate KJ, et al 'Impact of Dust from Mineral Workings' Mineral Planning into the 1990s, County Planning Officers Society, Committee No. 3 Conference, Loughborough University 19–21 September 1990

394. Wardell Armstrong 'Assessment of the Potential Impacts from Dust Generated during Opencast Mining and Quarrying' Leaflet of Wardell Armstrong Ltd, 24 August 1990

395. Pike DC 'Road Track Costs – Big Lorries versus Small' Transport Note TRN2, Sand and Gravel Association Ltd, 1976

396. Foster Yeoman Ltd 'Planning Application for an Extension to Torr Works' Foster Yeoman Ltd 1990

397. DOSCO 'Transportation System – Pipe Conveyor' Leaflet produced by Dosco Overseas Engineering Ltd, Tuxford, UK, 1990

398. Tompsett KR 'siteNoise – Open site noise prediction system' New Products, *Acoustics Bulletin* July 1990 p40

399. Anon 'Covered loads help reduce nuisance' First Sight, *Construction Weekly*, 5 September 1990

400. Hertfordshire County Council 'Minerals Criteria for the Assessment of Planning Applications' Hertfordshire CC June 1990

401. Elkington J 'The Environmental Audit – A Green Filter for Company Policies, Plants, Processes and Products' Sustainability Ltd London, 1990

402. The Royal Society for Nature Conservation 'The Peat Report' RSNC c1990

403. Ward E 'Levels of Vibration and Air Blast at Warmsworth Quarry' Blasting, *Mineral Planning* **44** September 1990 pp15–18

404. Arrowsmith R 'Quarry Face Profiling' *Mineral Planning* **44** September 1990 pp 18–20

18. References (Numerical)

405. Rutherford LA 'The Environmental Costs of Opencast Mining: Issues of Liability' *J of Environmental Law*, **2** 2, 1990, pp161–177

406. Kilpatrick FA et al 'Coal Hydrology Program of the US Geological Survey' Proceedings of Symposium on Surface Mining Hydrology, Sedimentology and Reclamation, Univ of Kentucky, Lexington US December 1980, pp113–117

407. Rosso WA 'Water: Resource or Liability in Surface Mining' Proc Conference & Expo 'Coal Age', Symposium on Surface Coal Mining & Reclamation, Louisville US October 1979, McGraw & Hill, New York 1979, pp99–102

408. Birch WJ et al 'Prediction of ground vibrations from blasting on opencast sites' *Trans Institution of Mining and Metallurgy* (Section A: Mining industry) **92**, April 1983, ppA103–107,

409. Walker AW 'Noise Standards – Where are they?' County Planning Officers Society, Committee no. 3, Conf 'Mineral Planning into the 1990s' Loughborough Univ September 1990

410. Siskind DE et al 'Comparative Study of Blasting Vibrations From Indiana Surface Coal Mines' US Bureau of Mines Report of Investigations RI9226, Washington, 1989

411. Siskind DE et al 'Structural Response and Cosmetic Cracking in Residences from Surface Mine Blasting' SANDIA Report, 59th Shock and Vibration Symposium **1**, October 1988, pp319–334

412. Proctor R 'Putting our energy into nature' British Coal Opencast Executive Publication c1990

413. Anon 'The Shropshire Badger Group and the British Coal Opencast Executive' Leaflet announcing a joint research project, published by BCOE, c1989

414. Department of the Environment/Welsh Office 'Mineral Planning Guidance: General Considerations and the Development Plan System', MPG1, HMSO January 1988

415. Siskind DE et al 'Structure Response and Damage Produced by Airblast from Surface Mining' US Bureau of Mines Report of Investigations RI8485, Washington, 1980

416. Siskind DE et al 'Structure Response and Damage Produced by Ground Vibration From Surface Mine Blasting' US Bureau of Mines Report of Investigations RI 8507, Washington, 1980

417. Olson KS et al 'Fugitive Dust Control for Haulage Roads and Tailing Basins US Bureau of Mines Report of Investigations RI9069, Washington, 1987

418. Kelsh DJ et al 'Dewatering Fine Particle Waste Suspensions with Direct Current' Chapter 27 *Encyclopedia of Fluid Mechanics*, Gulf Publishing Company, Houston Texas 1986, pp1171–1190

419. Sheiner BJ 'Increasing Dewatering Efficiency through Optmization of the Flocculation Sequence' Proceedings of International Technical Conference on 'Filtration and Separation' Ocean City, US March 1988, ed Frederick E R et al, pp371–378

420. Scheiner BJ 'New Dewatering Technique for Fine Particle Waste' *XVI International Mineral Processing Congress*, ed E Forssberg, Elsevier Science Publishers BV, Amsterdam 1988, pp1951–1961

421. Anon 'TEOM Series 1400, Ambient Particulate Monitor' Leaflet, Enviro Technology Gloucester c1990

422. HM Alkali and Clean Air Inspectorate 'Mineral works (roadstone plants) – Notes on Best Practicable Means' Health and Safety Executive BPM13, April 1982 pp1–4

423. Department of the Environment/ Welsh Office 'The use of conditions in planning permissions' Circular DOE/WO 1/85, HMSO 1985

424. Somerset CC 'Ditches and Rhynes – A Guide to Ditch and Rhyne Management for Wildlife' Somerset Levels and Moors, Guidelines No 1, Somerset County Council 1986

425. Confidential 'Radar Collision Prevention Device Trials' confidential communication 1990

426. Leicestershire CC 'British Gypsum Ltd' Decision Letter 87/1467/2, Leicestershire County Council, 18th June 1987

427. South Bucks DC, private communication, February 1990

428. Blue Circle Industries plc, private communication, January 1990

429. Buckinghamshire CC, private communication, November 1989

430. AF Budge (Mining) Limited, private communication, December 1989

431. Cleveland CC, private communication, February 1990

432. Clouston, private communication, November 1989

433. Dyfed CC, private communication, October 1989

434. Dyfed CC, Pembrokeshire Coast National Park, private communication, November 1989

435. Humberside CC, private communication, November 1989

436. Oxford Archaeological Associates Ltd, private communication, January 1990

437. Oxford Archaeological Unit, private communication, January 1990

438. Rock Environmental Ltd, private communication, July 1991

439. Suffolk CC, private communication, November 1989

440. Surrey CC, private communications, November 1989, January & May 1991

441. R & A Young Mining Limited, private communications, January & May 1990

442. National Rivers Authority, Welsh Region, private communication, March 1990

443. National Rivers Authority, Thames Region, private communications, March 1990, May 1991

444. National Rivers Authority, South West Region, private communication, March 1990

445. National Rivers Authority, Anglian Region, private communication, March 1990

446. National Rivers Authority, Severn-Trent Region, private communication, March 1990

447. National Rivers Authority, Wessex Region, private communication, March 1990

448. National Rivers Authority, Yorkshire Region, private communication, March 1990

449. The Royal Society for the Protection of Birds, confidential communication, March 1990

18. References (Numerical)

450. English Heritage, private communication, January 1990

451. Community Technical Aid – Northern Ireland, private communication, April 1990

452. Strathclyde RC, private communication, March 1990

453. West Glamorgan CC, letter and documents relating to Derlwyn Opencast public inquiry, 22nd May 1990

454. South Glamorgan CC, letter and extracts from the Council's evidence at a public inquiry into Rhoose Quarry, 9th May 1990

455. Somerset CC, letter 10th October 1990

456. British Coal Opencast Executive, letters of 6th July and 5th September 1990, 30th April 1991

457. Blue Circle Industries plc, private communications, July 1990, January & May 1991

458. Health & Safety Executive, private communication, July 1990

459. Health & Safety Executive, private communications, undated 1990, May 1991

460. Brett Gravel Ltd on behalf of the Sand and Gravel Association, private communications, July 1990 & May 1991

461. Nature Conservancy Council, private communications, July 1990 & January 1991

462. Walsall MBC, private communication, July 1990

463. Association of District Councils, private communications, July 1990 & January 1991

464. British Aggregate Construction Materials Industries, private communications, June & August 1990, January & May 1991

465. Redbridge LB, private communication, January 1990

466. National Federation of Clay Industries Ltd, private communications, August 1990, January & May 1991

467. Peak National Park, private communication, September 1990

468. Anon 'Government affirms belief in quality planning' *Planning* 896, 23rd November 1990, pp2–3

469. Anon, 'Government tightens up advice on archaeology' *Planning* 30th November 1990, p5

470. Anon, 'Brick firm tries new angle on odour problem' *Surveyor* 5 April 1984, p4

471. Anon, 'London Brick faces environmental objections to its Bedfordshire brickworks' *ENDS Report* 39 December 1979, pp13–18

472. Johnson, Paul & Bloomer, 'Application to extract brick clay from land adjacent to Oak Farm and restore to woodlands and openspace' support statement, ref GSW/HKS/89229-09, 18th April 1990

473. FAW Electronics Ltd, 'RAW490 Belltone Alarm' leaflet 1990

474. British Coal Opencast Executive, 'St Aidans Remainder – proposed opencast site' parts of planning application; document 2 'Environmental Statement' & document 3 'Non Technical Summary of Environmental Statement', BCOE, 1989

475. Staffordshire CC, 'Extraction of sand and gravel at Alrewas North, Alrewas, Redlands Aggregates Ltd' report to Planning Committee A.19(L) relating to application of 21.12.89, 1990 pp74–80

476. R & A Young Mining Ltd, 'Proposed Coal Mine at Whitwell' planning application (incorporating Environmental Statement), Application and Accompanying Statement 1, 1990

477. British Coal Opencast Executive, 'Coalfield West, Environmental Statement' Planning Application BCOE 1990

478. Anon, 'Draft "Duty of Care" Code soft on major waste producers' *ENDS Report* 181, February 1990, pp29–30

479. Steering Group and Working Group appointed by the Minister of Transport 'Traffic in Towns – A study of the long term problems of traffic in urban areas' HMSO London 1963

480. Coppin NJ et al, ed *Use of Vegetation in Civil Engineering*, B10 CIRIA/Butterworth 1990

481. Helliwell R, 'Moving Semi-Natural Vegetation' *Landscape Design* December/January 1989/1990 pp36–38

482. Worthington TR et al 'Transference of Semi-Natural Grassland and Marshland onto Newly Created Landfill' *Biological Conservation* 41 1987, pp301–311

483. Dowding CH 'Blast Vibration Monitoring and Control' Prentice-Hall NJ USA 1985

484. Anon, 'Scottish superquarry proposals – I'll take the high road' *Mineral Planning* 47 June 1991 pp22–23

485. Swain C, 'Mineral Policies in Development Plans' *Mineral Planning* 45 December 1990 pp7–12

486. Arup Economic Consultants, 'Mineral Policies in Development Plans' HMSO 1990

487. Anon, 'Aggregates Demand Attention' *Mineral Planning* 47 June 1991 pp15

488. Wilson CI, 'Scottish Superquarries – A catalyst for integrated development?' *Mineral Planning* 47 June 1991 pp18–21

489. Anon, 'Limestone Working Approved in AONB' *Mineral Planning* 45 December 1990 pp5–7

490. The Green Alliance, 'The Environmental Protection Act 1990' Green Alliance, London 1990

491. Department of the Environment/The Scottish Office/Welsh Office, 'Environmental Protection Act 1990, Part I Secretary of State's Guidance – Coal, coke and coal product processes', PG3/5(91), HMSO July 1991

492. Department of the Environment/The Scottish Office/Welsh Office, 'Environmental Protection Act 1990, Part I Secretary of State's Guidance – Quarry processes including roadstone plants and the size reduction of bricks, tiles and concrete', PG3/8(91), HMSO July 1991

493. Wilson K, 'A Guide to the reclamation of mineral workings to Forestry' R & D Paper 141, Forestry Commission 1985

494. Gagen PJ et al, 'A Geomorphological Approach to Limestone Quarry Restoration' *Geomorphology in Environmental Planning* ed Hooke J Wiley, New York 1988, pp121–142

CHAPTER 19

REFERENCES (Alphabetical)

Ackroyd T, 'High Hopes for Low Profile' *Mineral Planning*, 3 June 1980 pp1,8,9

Adkins N, 'Aids to Blast Design' *Quarry Management* October 1986, pp35-38

AF Budge (Mining) Limited, private communication, December 1989

Alabaster JS 'River Pollution by China Clay Wastes and Solids in Suspension' *Environmental Management of Mineral Wastes*, ed Goodman GT et al, Sijthoff & Moordhoff, Netherlands 1978 pp237-238

Allett EJ, 'Bretby and the Environment', *Colliery Guardian* March 1989, pp84-85

Ancich EJ, 'The Environmental Aspects of Structural Response to Blasting Overpressure', *Quarry Management and Products* July 1983, pp420-425

Andrews J, 'When You're Ready to Get Involved with About Face, What Do You Do?', *Stone Review*, June 1987, pp8-9

Anon, 'Clay waste mountain more than a molehill' *Construction News*, June 28th 1990, Special Report Quarrying and Mineral Extraction, p32.

Anon, 'Digging in for growth' *Construction News* June 28th 1990 Special Report Quarrying and Mineral Extraction p28

Anon, 'Thames Water makes a splash' *Construction News* June 28th 1990 Special Report Quarrying and Mineral Extraction p34.

Anon, 'C-Vision has safety in view' *Construction News*, June 28th 1990 Special Report, Quarrying and Mineral Extraction p34.

Anon, 'RWA 480 Electronic Reverse Warning Alarm' Data sheet 004 of FAW Electronics Ltd, Chesterfield, undated

Anon, 'Aerial ropeways – An efficient and economic transport system' *Quarry Management and Products* December 1981 pp 844-847.

Anon, 'How the environment was improved at Markle Mains Quarry' source unknown, March 1981 p9.

Anon, 'The Influence of Quarrying on Groundwater Resources' *Quarry Management and Products*, April 1979 pp97-99.

Anon, 'New Wimpey Plan to Secure Jobs and Improve Environment' *Quarry Management* July 1990 p37.

Anon, 'The Stability of Excavated Slopes at Opencast Coal Sites' Code of Practice published by the Federation of Civil Engineering Contractors, October 1989

Anon, 'The Siting and Construction of Temporary Spoil Mounds at Opencast Coal Sites' Code of Practice published by the Federation of Civil Engineering Contractors, reissued 1989

Anon, 'Code of Good Practice – Ground and Airborne Vibration From the Use of Explosives', Publication, Federation of European Explosives Manufacturers, Brussels 1985, pp21

Anon, 'Burlington Slate Invest in New Production Methods – 'Green' techniques improve yield' *Quarry Management* July 1990 p35.

Anon, 'Lorry Sheeting' *Mineral Planning* June 1990 pp12-14

Anon, 'The Original Smart Alarm Reversing Bleeper' Leaflet-/brochure published by Brigade Electronics Ltd, London, undated

Anon, 'Steetley Complete Plantation Removal – Rare species preserved in Thristlington project' *Quarry Management* August 1990 p31

Anon, 'Opencast Coal Productivity Maintained Despite Environmental Restrictions' *Quarry & Mining News*, No.96 January 18th 1980, p1

Anon, 'Demand Forecasts for Aggregates: "Nightmare Scenario" or Symbol of Prosperity?' *ENDS Report* **196** May 1991 pp 14-18

Anon, 'Legislation – Planning Gain and Agreements' *Mineral Planning* **40** September 1989 pp23-24

Anon 'The Shropshire Badger Group and the British Coal Opencast Executive' Leaflet announcing a joint research project, published by BCOE, c1989

Anon 'TEOM Series 1400, Ambient Particulate Monitor' Leaflet, Enviro Technology Gloucester c1990

Anon 'Government affirms belief in quality planning' *Planning* **896**, 23rd November 1990, pp2-3

Anon, 'Bigger Private Opencast Coal Mines face New Environmental Duty' *ENDS Report* **182** March 1990 p27

Anon, 'An Award Winning Restoration Scheme – Horton Trout Fishery – Haul Aggregates' *Construction News* June 28th 1990, Special Report on Quarrying and Mineral Extraction p26

Anon, 'Government tightens up advice on archaeology' *Planning* 30th November 1990, p5

Anon, 'Brick firm tries new angle on odour problem' *Surveyor* 5 April 1984, p4

Anon, 'London Brick faces environmental objections to its Bedfordshire brickworks' *ENDS Report* **39** December 1979, pp13-18

19. References (Alphabetical)

Anon, 'Draft "Duty of Care" Code soft on major waste producers' ENDS Report 181, February 1990, pp29-30

Anon, 'Scottish superquarry proposals – I'll take the high road' Mineral Planning 47 June 1991 pp22-23

Anon, 'Aggregates Demand Attention' Mineral Planning 47 June 1991 pp15

Anon, 'Limestone Working Approved in AONB' Mineral Planning 45 December 1990 pp5-7

Anon, 'Proposed Opencast Mining at Potatopot', Material for Exhibition, Workington 1982

Anon, 'Recent mineral cases', Mineral Planning 38 March 1989, pp14-18

Anon, 'Opencast Failure' Mineral Planning 38 March 1989, pp28-29

Anon, 'Brightsite', Campaign Information Pack, Groundwork, undated

Anon 'Covered loads help reduce nuisance' First Sight, Construction Weekly, 5 September 1990

Anon, 'Tarmac Railhead in Kent – Automated control systems at Hothfield' Quarry Management and Products, June 1981, pp403-408.

Anon, 'First Draft Guidance on Integrated Pollution Control' ENDS Report 188 September 1990 pp27-28

Anon, 'Success of US About Face Programme', Quarry Management, April 1988, pp45-46

Anon, 'Dust Control', Mining Magazine, June 1979, pp504-521

Anon, 'Quarry Planning: Is There a Role for Computers?' Mine & Quarry May 1990 pp19-23

Anon, 'Brussels takes a First Step towards an Environmental Auditing Directive' ENDS Report 192 January 1991 pp13-16

Anon, 'World Lead Claimed for British Standard on Environmental Management' ENDS Report 197 June 1991 p4

Anon, 'Emerging Issues in Environmental Auditing' ENDS Report 185 June 1990 pp 11-13

Anon, 'Environmental Management at English China Clays', M & Q Environment, 1 2, 1987, pp16-18

Anon, 'England's garden is handled with care' Construction Weekly, 12 June 1991, pp18-19

Anon, 'European Council of Ministers, Directive on the protection of Workers from Noise' Noise and Vibration Control Worldwide June 1986, p164

Anon, 'A Clean Sweep for Blue Circle', Quarry Management, April 1987, p40

Anon, 'Dust disposal at Halton East Quarry', Quarry Management 1987

Anon, 'Pont Einion and the Environment', Colliery Guardian July 1988, pp226-7

Anon, 'Who is Saying What About Mineral Planning', Bodytalk, Minerals Planning 40, September 1989, pp35-36

Anon, 'Tidying up the Mess of Britain's Discharge Consent System' ENDS Report 186 July 1990 pp17-24

Anon, 'Ffos Las opencast coal mine, South Wales', Mining Magazine, January 1987, p8

Anon, 'Time to Speak Out' Mining Magazine June 1990

Anon, 'National Parks: Review and Minerals Policies' Mineral Planning 44 September 1990 pp3-10

Anon, 'The use of screens to reduce noise from sites', CIRIA Practice Note, Special Publication 38, London, 1985

Anon, 'Crushing reply to problem of dump trucks' Quarrying, Construction News Supplement, 6 June 1991

Anon, 'Ancient Monuments in the Countryside', English Heritage Publication 1987, pp15

Anon, 'Commission Changes Tack on Environmental Auditing' ENDS Report 194 March 1991 p33

Anon, 'Effects of Particulate Air pollutants on Vegetation', Report of Subcommittee on Airborne Particles, Committee on Medical and Biologic Effects of Environmental Pollutants, Division of Medical Sciences, Assembly of Life Sciences, National Research Council; University Park Press, Baltimore Ch9 Airborne Particles 1979, pp199-214

Anon, 'Applications of Remote Sensing to Environmental Aspects of Surface Mining', Surface Mining, EOSAT Landsat Applications Issue 2, NRSC Farnborough, undated

Anon, 'Britain – DoE's Near-final List of Industries to be Controlled by Green Bill', ENDS Report 180 January 1990, p27

Anon, 'Restoration Bonds put under the Microscope' Planning, 19 April 1991

Anon, 'Go ahead for Cheltenham extraction/landfill site', Planning News, M & Q Environment 3 2, 1989, p6

Anon, 'The environmental impact of mining and quarrying', ENDS Report 21 March 1979, pp13-15

Anon, 'No 'Mappa Minerals' in Lugg Valley', Recent mineral cases, Mineral Planning 40, September 1989, pp27-28

Anon, 'Woolhampton Quarry – An example of rapid and environmentally sensitive development', M & Q Environment 2 5, 1988, pp26-28

Anon, 'Mendip Limestone: And then there was one...' Mineral Planning 47 June 1991 p17

Anon, 'New Tarmac Quarry in Somerset', Quarry Management, March 1986, pp13-18

Anon, 'Solving Environmental Problems' World Mining Equipment 9 6, June 1985, pp13-16

Anon, 'Stateside Stripping – Taywood Mining's environmental experience in America', M & Q Environment 1 2, 1987, pp22-24

Anon, 'The About Face Program: More Than Just Beautification', Stone Review April 1987, pp8-19

Anthony Goss Associates et al, 'Planning Conditions for Hard Rock Quarries' Report for the DoE 1983

ARC Ltd for Stone Federation, private communication July 1991

Arrowsmith R 'Quarry Face Profiling' Mineral Planning 44 September 1990 pp 18-20

Arup Economic Consultants, 'Mineral Policies in Development Plans' HMSO 1990

19. References (Alphabetical)

Association of District Councils, private communications, July 1990 & January 1991

Atkins Research and Development, 'Environmental Impact of Mining and Quarrying Activities', Monograph December 1980

Atkinson T et al, 'Surface Mining - Future Concepts', Proceedings of symposium 'Surface Mining -Future Concepts' University of Nottingham, 18-20th April 1989, pp5-11

Atlee F, 'Wake Stone: Exemplifying the Good Neighbour Policy and Reaping the Rewards ', *Stone Review*, April 1987, pp10-13

Attewell PB, 'Noise and vibration in civil engineering', *Municipal Engineer* 3 June 1986, pp139-158

Avon, Gloucestershire, and Somerset Environmental Monitoring Committee, 'Vibration. A guide for Environmental Health Officers', Publication available from Hon Secretary, Environmental Health Department, Wansdyke District Council, Midsomer Norton, June 1985

Badham IF, 'Mining and Environmental Noise - Preventing Problems Professionally', Australian IMM Symposium, Mining and Environment - A Professional Approach, Brisbane July 1987, pp109-115

Ball M, 'A Review of Blast Design Considerations in Quarrying and Opencast Mining - Part 1 Principles of design, Part 2 Initiation and computerized design', *Quarry Management* June 1988, pp35-39 and July 1988, pp23-27

Ball MJ et al, 'Field Experience with new Methods of Electric Shotfiring' 8th Annual Conference of the Society of Explosives Engineers, New Orleans, February 1982

Barnett JL et al, 'Fugitive Emissions - a Directional Sampler for Particulates', Warren Spring Laboratory report published by Department of Trade and Industry, Stevenage, 1987

Bate RR, 'Opencast Coalmining: The view of the CPRE' IMM Symposium, Minescape '88, 1988, pp171-177

Bate KJ, et al 'Impact of Dust from Mineral Workings' Mineral Planning into the 1990s, County Planning Officers Society, Committee No. 3 Conference, Lougborough University 19-21 September 1990

Bate RR et al, 'Campaigners' Guide to Opencast Mining' CPRE February 1991

BCOE, 'Complaint Report', BCOE 1990

BCOE, 'St Aidans Remainder - proposed opencast site' parts of planning application; document 2 'Environmental Statement' & document 3 'Non Technical Summary of Environmental Statement', BCOE, 1989

BCOE, letters of 6th July and 5th September 1990, 30th April 1991

BCOE, 'Coalfield West, Environmental Statement' Planning Application BCOE 1990

BCOE, 'Complaints in 1989' Internal Memorandum 12 March 1990

BCOE, 'Complaints in 1990' Internal Memorandum 26 February 1991

BCOE, 'Opencast Coal Mining in Great Britain' BC Booklet 1988, pp 22

Beaman AL et al 'Recent Developments in the Method of Using Sticky Pads for the Measurement of Particulate Nuisance' *Clean Air* 14 2, 1984, pp74-81

Beaman AL et al 'Assessment of Nuisance from Deposited Particulates Using a Simple and Inexpensive Measuring System' *Clean Air* 11 2, 1981, pp 77-81

Beaman AL 'NTPC Ash Disposal - Review of Soil Stabilisers' Report by W S Atkins & Ptnrs September 1988

Beeley JG, 'Instrumentation and the Surface Environment', IMM Symposium, Minescape '88, 1988, pp53-67

Bennett JH et al, 'Assessment of Fugitive Emissions from Sand and Gravel Processing Operations', 73rd Annual Meeting of the Air Pollution Control Association, Montreal June 22-27, 1980, pp14

Birch WJ et al 'Prediction of ground vibrations from blasting on opencast sites' *Trans Institution of Mining and Metallurgy* (Section A: Mining industry) 92, April 1983, ppA103-107,

Blue Circle Industries plc, private communication, January 1990

Blue Circle Industries plc, private communications, July 1990, January & May 1991

Blue Circle Industries plc, 'Hope', Proposal produced by Blue Circle Industries 1988 relating to Hope Cement Works 1988

Bollinger GA, 'Blast Vibration Analysis' Southern Illinois University Press, Carbondale, 1971.

Bowhill A, 'Balancing Nuisance with Need' *Land and Minerals Surveyor* 7 1989, pp535-537

Brandt CJ et al, 'Effects of Limestone Dust Accumulation on Composition of a Forest Community' *Environmental Pollution* 3 1972 pp217-225

Brandt CJ et al, 'Effects of Limestone Dust Accumulation on Lateral Growth of Forest Trees' *Environmental Pollution* 4 1973 pp207-213

Brent-Jones E, 'Opencast Coal Mining in the UK', RICS Conference 22nd October 1981, pp21

Brett Gravel Ltd on behalf of the Sand and Gravel Association, private communications, July 1990 & May 1991

Bristow C, 'How New Patterns of Demand for Minerals are Created by Changing Socio-economic Trends' Science 89, Annual Meeting of British Association, Sheffield September 1989 Paper No GI.12

British Standards Institution, 'Environmental Management Systems' Draft British Standard, EPC/50, Documents 91/53255-7, BSI 1991

British Standards Institution, 'Methods for the Measurement of Air Pollution', British Standard 1747:1972, BSI London 1972

British Materials Handling Board, 'Code of Practice for the Purchase and Operation of Fabric Filters for Dust Control', British Materials Handling Board August 1985.

British Aggregates and Construction Materials Industry, 'Why Quarry? - 6 - The Caring Face of Quarrying' A BACMI Information Sheet, London 1989

British Standards Institution, 'Guide To Evaluation of Human Exposure to Vibration in Buildings (1Hz to 80Hz)' Draft Standard (Revision of BS 6472 : 1984), BSI London, 3 February 1990

British Standards Institution, 'Noise control on construction and open sites', British Standard BS5228: Parts 1-3: BSI London 1984

19. References (Alphabetical)

British Standards Institution, 'Methods for the measurement of Air Pollution – Part 5 Directional Dust Gauges' BS1747:1972, BSI London, 1972

British Aggregate Construction Materials Industries, private communications, June & August 1990, January & May 1991

British Standards Institution, 'British Standard Guide to 'Evaluation of human exposure to vibration in buildings (1 Hz to 80 Hz)'', BS 6472:1984 BSI London, 1984

British Standards Institution, 'Method for Rating Industrial Noise Affecting Mixed Residential and Industrial Areas', British Standard 4142: 1990, BSI London 1990

Broadhurst KA et al, 'Blast Induced Ground Vibration and Air Overpressure– Terminology, Standards and Control', Paper published by Rock Environmental Ltd, undated

Broadhurst KA, 'Blast Vibration Evaluation', Paper published by Rock Environmental Limited, Derby, 1987

Broadhurst K et al, 'Review of Current Standards and Recommendations for Vibration and Noise', IMM Conf 14/2/1984

Brook C et al, 'Environmental Aspects of Opencast Mining with reference to the Rother Valley Country Park', *The Mining Engineer* November 1989, pp191–196

Brook C et al, 'Opencast Blasting and the Environment', *The Mining Engineer*, December 1988, pp253–8

Buckinghamshire CC, private communication, November 1989

Buckinghamshire CC, County Planning Officer, 'Report of the County Planning Officer', Planning Sub-committee 13th November 1989

Burnham DP et al, 'Developments in Blasting Practice', *Quarry Management and Products*, February 1982, pp95–102

Burnham DP et al, 'Developments in Blasting Practice' *Quarry Management & Products* October 1981

Canessa W, 'Dust Retardants ', Chapter 16 of 'Air Pollution Control and Design Handbook' Part 1 Edited by Cheremisinoff PN et al, Dekker 1977, pp431–447

Carden S, 'Opencast Mining – Complaints and Nuisance', *M & Q Environment* 3 2, 1989 p15

Carnwath R, 'Enforcing Planning Control' Report to the Department of the Environment, HMSO, Feb 1989 .

Carpenter SL, 'Environmental Conflicts in Mineral Development or The Art of Creating Unnecessary Problems' *Minerals and the Environment*, 2 1989 pp159–164.

CBI, 'Archaeological Remains' Chairman's letter to Members CBI Minerals Committee, 7th April 1982

CBI, 'Archaeological investigations – Code of Practice for Minerals Operators', CBI Publication March 1991

Chadwick MJ et al, *Environmental Impacts of Coal Mining and Utilisation Environmental Impacts of Coal Transportation* Pergamon Press 1987 Ch. 4

Chadwick MJ et al, *Environmental Impacts of Coal Mining and Utilisation Land Disturbance and Reclamation after Mining*, Pergamon Press 1987 Ch. 2

Chandler AJ et al, 'The Application and Verification of Open Source Fugitive Emission Estimates', paper 78-55.5, 71st Annual Meeting of the Air Pollution Control Association, Houston June 25-30 1978

Cheer J S, 'Berkshire County Council Appeal by S Grunden (EWELME) Ltd.' Report to Secretary of State for the Environment re extraction of sand at Old Kiln Farm, Chieveley, Newbury, ref DPI101/P6/ST/P October 1988.

CIRIA, 'Construction Industry Research – Rock-Blasting: A Guide to Good Practice' Document 1174/WP1/1, CIRIA September 1989

Civic Trust, County Surveyors Society, Department of Transport, 'Lorries in the community', HMSO 1990

Clark AR et al, 'Magnesian Limestone Quarrying Planning and Environmental Considerations', *M & Q Environment* 3 1, 1989, pp15–17

Clement K, 'Environmental Auditing for Industry: A European Perspective' *European Environment*, 1 3 1991, pp1–4

Cleveland CC, private communication, February 1990

Clouston, private communication, November 1989

Coal Information and Consultancy Services, 'The British Opencast Industry' Report CICS London, March 1989

Coalfield Communities Campaign, 'Opencast Coal', Report by the CCC Opencast Coal Working Party, 1987

Collens G, 'Coolscar 2: Landscape Issues', *Mineral Planning* 32 September 1987, pp26–9

Colley R et al, 'Reviewing the Quality of Environmental Statements' *The Planner* 27 April 1990, pp12–13

Commission on Energy and the Environment, 'Coal and the Environment' HMSO 1981, p257

Commission on Mining and the Environment, 'Report of the Commission on Mining and the Environment' HMSO 1972

Community Technical Aid – Northern Ireland, private communication, April 1990

Confidential 'Radar Collision Prevention Device Trials' confidential communication 1990

Contractors' Aggregates Ltd, 'Environmental Statement – Land at Chantry Farm/Toppinghoe Hall, Boreham, Essex', published by author, February 1989, pp32

Cookson T, 'Dartmoor National Park, Appeal by British Rail Property Board' Report to Secretary of State for the Environment re Meldon Quarry, Okehampton, Ref A/621X/JLC/P 22nd January 1988.

Coppin NJ et al, ed *Use of Vegetation in Civil Engineering*, B10 CIRIA/Butterworth 1990

Coppin N, et al 'Environmental Assessment for Opencast Coal Mining' *M & Q Environment* 3 3, 1989, pp 17–22

Corker SP, 'Bulk Soil Handling Techniques', *M & Q Environment*, 1 1, 1987, pp15–16

Cotton J, 'Archaeological Investigations at Holloway Lane 1980–88', pamphlet pub by Museum of London & Drinkwater Sabey 1988

Council for Environmental Conservation, 'Scar on the Landscape?', Report on Opencast Mining published by CoEnCo London, 1979.

Cowe R, 'Greening by numbers: a new role for accountants' *Guardian* 19 June 1990

Cross D, 'A Slippery Situation' *Surveyor* 1 March 1990

19. References (Alphabetical)

Cumbria CC & Lake District Special Planning Board, 'Cumbria & Lake District – Joint Minerals Local Plan (Excluding Coal)', Written Statement March 1986

Cumbria CC, 'Planning permission for Potatopot', Reference 2/84/0749, 8th June 1985, letter to NCBOE

Cumbria CC, 'Cumbria Coal Local Plan – Draft Written Statement for Consultation', September 1988

Cunningham AE, 'Coal Production and the Environment' Department of Energy Library Bibliography 1979

Dalgleish I, 'Cost-Effective Drilling and Blasting – Developments contribute to efficiency and safety', *Quarry Management*, January 1989, pp19–27

Davies IV et al, 'Opencast Coal Mining: working, restoration and reclamation' IMM Symposium, Minerals and the Environment, London, June 1974, pp313–331

Davies MR et al, 'An Application for Planning Permission to Extend Sand and Gravel Quarrying Operations at Lydd, Kent and Camber, East Sussex – 2 Environmental Statement, Publication by Brett, August 1989.

Denby B, 'Computer Aids in Quarry Design and Planning – Opencast techniques applied to quarrying', *Quarry Management* May 1988, pp27–41

Department of the Environment, 'Planning Policy Guidance: Archaeology and Planning' PPG16, HMSO November 1990

Department of the Environment et al, 'Coal and the Environment', Government's response to the Commission on Energy and the Environment's Report, HMSO London, 1983

Department of the Environment, MAFF and Welsh Office, 'The Water Act – Code of Practice on Conservation, Access and Recreation' DOE, July 1989, pp12–16

Department of the Environment/Welsh Office 'Mineral Planning Guidance: General Considerations and the Development Plan System' MPG1, HMSO January 1988

Department of the Environment/Welsh Office, 'Minerals Planning Guidance: Applications, Permissions and Conditions', MPG2, HMSO 1988

Department of the Environment/Welsh Office, 'Minerals Planning Guidance: Opencast coal mining', MPG3, HMSO, May 1988

Department of the Environment/Welsh Office, 'Mineral Planning Guidance: The review of Mineral Working Sites' MPG 4, HMSO September 1988.

Department of the Environment/Welsh Office, 'The Reclamation of Mineral Workings' MPG7, HMSO 1989

Department of the Environment/Welsh Office, 'Planning and Noise', Circular DOE 10/73, WO 16/73, HMSO 1973

Department of the Environment/Welsh Office, 'Environmental Assessment – A Guide to the Procedures' HMSO 1989

Department of the Environment/Welsh Office, 'Town and Country Planning Act 1971, Planning Gain' Circular DOE 22/83, WO 46/83, HMSO August 1983.

Department of the Environment/Welsh Office 'The use of conditions in planning permissions' Circular DOE/WO 1/85, HMSO 1985

Department of the Environment/Welsh Office, 'Public Rights of Way' Circular DOE/WO 1/83, HMSO February 1983

Department of the Environment/Welsh Office, 'EC Directive on the Protection of Groundwater Against Pollution Caused by Ceertain Dangerous Substances (80/68/EEC)' Circular DOE 4/82, WO 7/82, HMSO March 1982

Department of the Environment/Welsh Office, 'EC Directive on the Protection of Groundwater Against Pollution Caused by Certain Dangerous Substances (80/68/EEC): Classification of Listed Substances' Circular DOE 20/90, WO 34/90, HMSO March 1982

Department of the Environment/Welsh Office, 'Integrated Pollution Control' A Consultation Paper July 1988

Department of the Environment/MAFF, 'Conservation Guidelines for Water Authorities, Internal Drainage Boards, Local Authorities, The Nature Conservancy Council and the Countryside Commission', Chapter 7 Protection of Archaeological and Other Historic Features, MAFF/DoE Publication 1988, pp15–16

Department of the Environment/The Scottish Office/Welsh Office, 'Environmental Protection Act 1990, Part I Secretary of State's Guidance – Coal, coke and coal product processes', PG3/5(91), HMSO July 1991

Department of the Environment/The Scottish Office/Welsh Office, 'Environmental Protection Act 1990, Part I Secretary of State's Guidance – Quarry processes including roadstone plants and the size reduction of bricks, tiles and concrete', PG3/8(91), HMSO July 1991

Doncaster MBC Planning Directorate, 'Levels of Vibration and Air Blast at Warmsworth Quarry', Report 1990

Donnison RD, 'Staffordshire County Council, Appeal and Applications by The British Coal Corporation: Opencast Executive' Report to Secretary of State for the Environment on the Brown Lees opencast working, Ref A/139X/KP/P, 30 Nov 1987.

Donnison RD, 'Derbyshire County Council Appeal by British Coal Corporation (Opencast Executive)' Report of Inspector to Secretary of State for the Environment, ref APP/D1000/A/88/87968, 14th February 1989 (opencast coal and clay, Kirk & Ryefield)

DOSCO 'Transportation System – Pipe Conveyor' Leaflet produced by Dosco Overseas Engineering Ltd, Tuxford, UK, 1990

Dowding CH 'Blast Vibration Monitoring and Control' Prentice-Hall NJ USA 1985

Dowding CH et al, 'Simultaneous Airblast and Ground Motion Response' *Journal of the Structural Division, Proceedings of ASCE* **108** ST11, November 1982, pp2363–78

Down CG et al, *'Environmental Impact of Mining'* Ch 4 'Air Pollution', Applied Science Publishers 1977, pp57–88

Down CG et al, *'Environmental Impact of Mining'* Ch 3 'Visual Impact', Applied Science Publishers 1977, pp28–56

Down CG et al, *'Environmental Impact of Mining'* Ch 5 'Water Pollution', Applied Science Publishers 1977, pp89–144

Down CG et al, 'The Environmental Impact of Large Stone Quarries and Open-Pit Non-Ferrous Metal Mines in Britain', Final report of the Mining Environmental Research Unit RSM, Imperial College June 1976

Down CG et al, *'Environmental Impact of Mining'* Ch 6 'Noise' Applied Science Publishers 1977, pp142–163

Down CG et al, *'Environmental Impact of Mining'* Ch 2 'Range and Importance of Environmental Problems', Applied Science Publishers 1977, pp10–27

19. References (Alphabetical)

Down CG, 'The Design and Construction of Amenity Banks and Tips – with particular reference to landscaping', Report to Department of the Environment, Royal School of Mines, September 1976, pp243

Dumenil P, 'The use of explosives in quarries' Blue Circle notes for Meeting Institute of Quarrying, 9th November 1989

Dunn P 'Blasting and Vibration control in British Opencast Mining' *Mine and Quarry* Jan/Feb 1989 pp59–63

Dyfed CC, Pembrokeshire Coast National Park, private communication, November 1989

Dyfed CC, private communication, October 1989

Dyson S, 'Quarrying takes flight' *Construction News*, June 28th 1990 Special Report, Quarrying and Mineral Extraction p30.

EC, 'Council Directive on the protection of groundwater against pollution caused by certain dangerous substances' Official J of the European Communities No. L 20/43–47, 26 January 1980

EC, 'on The Assessment of the Effects of Certain Public and Private Projects on the Environment' Council Directive 85/337-/EEC, 27 June 1985.

EC, 'Draft Proposal for a Council Regulation Establishing a Community Environmental [Management] Auditing Scheme (Eco-Audit)' Document No. XI/83/91 – Rev 2, CEE Brussels May 1991

EC, 'A Consultation Paper on Draft Elements for a Council Directive on the Environmental Auditing of Certain Industrial Activities' EEC, XI/632/90–EN December 1990

EC, 'Proposal for a Council Directive on the Conservation of Natural and Semi-Natural Habitats and of Wild Fauna and Flora' Restricted Document 5949/91, Brussels 3 May 1991

Elkington J 'The Environmental Audit – A Green Filter for Company Policies, Plants, Processes and Products' Sustainability Ltd London, 1990

Ellis AF, 'The Control of Air Pollution in the Minerals Industry', Proceedings of 49th NSCA Annual Conference Llandudno 19th October 1982, NSCA Brighton 1982

Ellis DM et al, 'A Candle in the Wind?' *Mineral Planning* **38** March 1989, pp3–13

English Heritage, private communication, January 1990

Etherington JR, 'The Effect of Limestone Dust on a Limestone Heath in South Wales', *Nature in Wales* **15**, 1977, pp218–223

FAW Electronics Ltd, 'RAW490 Belltone Alarm' leaflet 1990

Finlayson D, 'Ballymoney Lignite Project – Hydrology and Environmental Assessment', *M & Q Environment* **3** 2, 1989, pp17–19

Finley BJ 'An Application for Planning Permission to Extend Sand Quarrying Operations into Foxes Bank – Ryarsh, Environment Assessment **2**' Published by Ryarsh Brick Ltd, March 1990

Fish B, 'Quarrying and the Community' *Stone Review*, June 1987 pp6–7

Fish BG, 'Environmental and amenity issues: significance and priorities' *Quarry Management* August 1983, pp487–492

Fleming BF, 'A Study of the Effects of Quarries on Adjacent Woodland', Report for MSc, Centre for Environmental Technology, Imperial College, London, 1986, pp124

Fletcher LR et al, 'Reducing accidents through improved blasting safety', Part of Information Circular 9135, 'Surface Mine Blasting', Proceedings: Bureau of Mines Technology Transfer Seminar, Chicago, 1987, pp6–14+

Forestry Commission, 'Catalogue of Publications' Forestry Commission, Farnham 1991

Foster Yeoman Ltd 'Planning Application for an Extension to Torr Works' Foster Yeoman Ltd 1990

Froedge DT, 'Outsmarting Blast Vibrations' *Coal* November 1989, pp67–69

Fuchs P, 'To quarry or not to quarry... Blot on landscape', *Construction News*, January 11th 1990, pp16–19

Gagen PJ et al, 'A Geomorphological Approach to Limestone Quarry Restoration' *Geomorphology in Environmental Planning* ed Hooke J, Wiley, New York 1988, pp121–142

Garner K, 'Controlling the Quarry Dust Hazard' *Quarry Management and Products* November 1983 p753

Geoffrey Walton, 'Review of current practice in British quarries etc.' Tasks 2&3 of DoE commissioned report, February 1986

Geoffrey Walton, 'Handbook on the hydrology and stability of excavated slopes in quarries', Task 5 of DoE commissioned report, 2nd impression 1988

Geoffrey Walton, 'The stability and hydrogeology of mineral workings', Summary of findings of previous tasks commissioned by Department of the Environment, July 1986

Gilbert OL, 'An Alkaline Dust Effect on Epiphytic Lichens', *Lichenologist*, **8** 1976, pp173–178

GLC, 'Beddington Farm Lands – Draft Planning Brief for Gravel Extraction and Restoration', GLC, BMA05, 1983

Glover HG, 'Mine Water Pollution – an Overview of Problems and Control Strategies in the United Kingdom' *Water Science and Technology* **15** 2, 1983, pp59–70

Golder Associates (UK) Ltd, 'Blast Diagnosis Service', Brochure Published by Golder Associates (UK) Ltd February 1989

Gosling D, 'Exposing Some of the Myths about Opencasting – A Case Study' *Mineral Planning*, March 1990 pp8–10
Wilton T J, 'Marley Hill Opencast Site' Proof of Evidence Public Enquiry, Rock Environmental Ltd, 31st October 1989

Gowan M, 'Water in the China Clay Industry' *Journal IWEM* **1** 1, August 1987, pp123–128

Griffin M J et al, 'Discomfort produced by impulsive whole-body vibration' *Journal of Acoustical Society of America* **68** 5, November 1980, pp1277–1284

Griffin CG, 'Ground Vibration and Air Overpressure from Blasting', Guidance Note, Cornwall County Council, July 1989

Griffiths G, 'Past Perfect? The DoE Planning Policy Guidance on Archaeology and Planning' *Mineral Planning* **46** March 1991 pp9–12

Grigg CFJ, 'Landscaping Techniques and Restoration', *Mining Magazine*, December 1988, pp492–497

Grimshaw PN, 'Environmental Benefits of Surface Mining' *Mining Magazine* December 1986 pp581–5

Grimshaw PN, 'Butterwell Mine, UK has much to Celebrate', *Mining Magazine*, January 1987, pp28–37

19. References (Alphabetical)

Groom AR, 'Planning Conditions and the Quarry Manager' *Quarry Management and Products* July 1982, pp5

Groundwork Associates Ltd, 'The Minerals Industry Environmental Performance Study' Groundwork Associates Ltd, 1991

Hall DJ et al, 'An Improved, Readily Available Dustfall Gauge' *Atmospheric Environment* **20** 1 pp219-222

Hallett S et al, 'UK Water Pollution Control: A Review of Legislation and Practice' *European Environment* **1** 3 1991, pp7-13

Hammersley R, 'Opencast Coal: A genuine cause for concern?' *Mineral Planning* March 1990 pp5-8

Hampshire CC, Various Reports by County Planning Officer to the Roads and Development Sub-Committee of Hampshire CC re applications for planning permission for minerals operations 1988-9

Hampshire CC, "Model' planning conditions', Publication 1/1988, Hampshire CC, January 1988

Hancock RP, 'Visual Response to Dustiness', *Journal of the Air Pollution Control Association*, January 1986 **26** 1, 1986, pp54-57

Hantz D et al, 'Principaux effets de l'exploitation des carrières sur l'environnement immédiat', *Mines et Carrières* **71**, May 1989, pp46-50

Hardaway J 'Coal Mining and Ground Water' *Surface Coal Mining Technology - Engineering and Environmental Aspects* ed Fung R, Noyes Data Corporation, NJ USA, 1981, pp 350-361

Hawkins MR, 'The review of Mineral Working Sites; Results of Questionnaire Survey - Priorities for Action' Report of the County Planning Engineer and Planning Officer to Planning Sub-Committee, Devon CC, EP/91/5/HQ 2 Jan 1991

Health & Safety Executive, private communications, undated 1990, May 1991

Health & Safety Executive, 'Explosives at Quarries - Quarries [Explosives] Regulations 1988 - Approved Code of Practice' HMSO, London, 1989

Health & Safety Executive, 'Fly Rock Projection from Quarry Blasting' Quarries Topic Report, Health and Safety Executive, September 1989

Health & Safety Executive, private communication, July 1990

Heasman IN, 'Mega-Quarries - The environmental impact of the trend towards giantism' *M & Q Environment* **3** 1, 1989, pp23-27

Helliwell R, 'Moving Semi-Natural Vegetation' *Landscape Design* December/January 1989/1990 pp36-38

Hertfordshire County Council 'Minerals Criteria for the Assessment of Planning Applications' Hertfordshire CC June 1990

Hibbard WR, 'Environmental Impact of Mining', *A Sourcebook on the Environment - a guide to the literature*, University of Chicago Press 1987, Ch 14 pp295-308

Higgins T, 'SAGA backs green agenda' *Construction News*, June 28th 1990 Special Report, Quarrying and Mineral Extraction p30.

Hill RD, 'The Impacts of Coal Mining on Surface Water and Control Measures Therefor' *Surface Coal Mining Technology, Engineering and Environmental Aspects*, ed Fung R, Noyes Data Corporation, New York, 1981, pp362-367

HM Alkali and Clean Air Inspectorate 'Mineral works (roadstone plants) - Notes on Best Practicable Means' Health and Safety Executive BPM13, April 1982 pp1-4

HM Inspectorate of Pollution, 'Best practicable means: general principles and practices' HMIP, BPM 1/88, January 1988

Hodges DJ et al, 'Computer graphics for public enquiries', *Colliery Guardian*, February 1986, pp56-62

Holmes PJ, 'The Environmental Impact of Quarrying - Planning procedures and mitigation measures in Australia', *Quarry Management*, April 1988, pp37-43

Hughes TEV (Inspector), 'Report on the Objections to the Plan made at the Public Local Inquiry [into Radyr/Morganstown District Plan]', Date of Inquiry October 1981, 1982

Humberside CC, private communication, November 1989

Hurrell M, 'Extracting industry needs to keeps its footing under exacting environmental demands' *Quarrying, Construction News Supplement*, 6 June 1991 pp4&6

ICI Explosives 'Blast Vibrations Frequency Amplitude Modelling' *Down Line* 12, ICI Explosives, February 1990, pp20-22

IEA Coal Research, 'Coal Research Projects 1988', Abstracts of current and recent technology R&D 1988, pp440-458

Anon, 'Why harm nature for peat's sake?' *Independent*, 11th December 1989

International Chamber of Commerce, 'Environmental Auditing' ICC position paper, Paris 1988

Jackson JR, 'Environmental Problems Associated with Opencasting - a Selective Review of Practical Considerations' *Clean Air* **11** 2, 1981, pp66-67

Jarvis D, 'Visual Impacts - Who Sees What from Where?', *Minerals Planning*, 26 March 1986, pp8-9

Jarvis D, 'The Use of Computers in Quarry Planning and Design', *M & Q Environment* **1** 1, 1987, pp 10-11

Jay Mineral Services Ltd, 'Mineral Planning Authority Blasting Vibrations Survey', Report by JMS, Bissoe, Truro, Cornwall TR4 8QZ, July 1988

Jenkins B, 'Integrating environmental factors into mining projects' Australian IMM Symposium, Mining and Environment - A Professional Approach, Brisbane July 1987 pp101-108

Jenkins B, 'Water and Waste Management for the Pine Creek Gold Mine', *Mine & Quarry* January/February 1987, pp61-62

Johnson D, 'New Tilcon Quarry Meets Environmental Standards', *Stone Review* February 1988 pp15-17

Johnson, Paul & Bloomer, 'Application to extract brick clay from land adjacent to Oak Farm and restore to woodlands and openspace' support statement, ref GSW/HKS/89229-09, 18th April 1990

Johnson D, 'New Tilcon Quarry Meets Environmental Standards', *Stone Review*, February 1988, pp15-17

Johnson FM, 'Aspects of Planning in Opencast Mining', From 'Surface Mining - Future Concepts' Proceedings of symposium University of Nottingham 18-20 April 89, pp30-34

Jones B, 'Open-cast mining blamed for levels of illness - Doctors say asthma cases 15 times average in town near coal workings', *The Times*, Monday June 6th 1988, p6

Just CD et al, ' The economic and operational implications of blast vibration limits' Australian IMM Symposium, Mining and Environment - A professional Approach, Brisbane July 1987, pp177-124

19. References (Alphabetical)

Kaas LM, 'Bureau of Mines Involvement in Environmental Issues' American Mining Congress, Mining Technology and Policy Issues 1986, Las Vegas October 1986

Kelsh DJ et al 'Dewatering Fine Particle Waste Suspensions with Direct Current' Chapter 27 *Encyclopedia of Fluid Mechanics*, Gulf Publishing Company, Houston Texas 1986, pp1171–1190

Kemmann-Lane T, 'Northumberland C C Appeal by British Coal Corporation' Report to the Secretary of State for the Environment regarding Linton Opencast Workings Ref D/1X/HB/P 6th September 1988.

Kemmann-Lane T, 'Peak Park Joint Planning Board, Appeal under Section 36 of the Town and Country Planning Act 1971 by T W Ward (Roadstone) Ltd. against the refusal of planning permission for the quarrying of limestone at Eldon Hill Quarry' Report of Inspector to the Secretary of State for the Environment, Reference A/557X/AJT/P, 17th December 1986.

Kerry G, 'Dealing with Quarry Blast Noise', Proceedings of Symposium 'The Problem of Assessing Environmental Noise', The Association of Consulting Scientists, London, 25th March 1981

Kerry G et al, 'Towards a Greater Understanding of Sound Levels from Blasting', *Quarry Management and Products*, April 1983, pp213–217

Kestner MO, 'Enclosure, wet suppression will help control dust', *Pit & Quarry* January 1988, pp72–74

Kestner MO, 'Ten guidelines for developing a plant's dust control system', *Pit & Quarry*, c1988

Ketteridge D, 'The Politics of Dust' *Mine and Quarry* Jan/Feb 1989, pp57–8

Kilpatrick FA et al 'Coal Hydrology Program of the US Geological Survey' Proceedings of Symposium on Surface Mining Hydrology, Sedimentology and Reclamation, Univ of Kentucky, Lexington US December 1980, pp113–117

Kopp JW, 'Initiation Timing Influence on Ground Vibration and Airblast', Information Circular 9135, 'Surface Mine Blasting', Proceedings: Bureau of Mines Technology Transfer Seminar, Chicago, 1987, pp51–59

Kopp JW, 'Stemming Ejection and Burden Movements of Small Borehole Blasts', Information Circular 9135, 'Surface Mine Blasting', Proceedings: Bureau of Mines Technology Transfer Seminar, Chicago, 1987, pp 106–114

Laage LW et al, 'Discriminating Back-up Alarm Technology for Front End Loaders', *The Mining Engineer* December 1988, pp276–277

Lambert AK, 'Dust Control at Roadstone Plants', *M & Q Environment* 1 2, 1987, pp14–15

Lane DD et al, 'Use of Laboratory Methods to Quantify Dust Suppressant Effectiveness', *Trans Society of Mining Eng of AIME* 274, pp2001–2004

Leicestershire CC 'British Gypsum Ltd' Decision Letter 87/1467/2, Leicestershire County Council, 18th June 1987

Leuschner H-J, 'The Opencast Mine of Hambach – A synthesis of raw material winning and landscaping', IXth World Mining Congress II-1, 1986, pp1–17

Libicki J, 'Changes in Groundwater due to Surface Mining', *Int Journal of Mine Water*, 1 1982, pp25–30

Lindorff DE, 'Hydrogeology of Surface Mine Spoil in Illinois' Annual Meeting Society of Mining Engineers of AIME, Illinois, Preprint 81-24 February 1981

Local Authority Associations, 'Opencast Mining – Code of Practice' LAA Code, June 1986

Machin AW, 'Cambridgeshire County Council Appeal by Tarmac Roadstone Ltd. Report of Inspector to Secretary of State' B/314/SM/P 1st December 1988 (Sand & Gravel, Bridge Farm, Willingham)

Martin JF, 'Coal Refuse Disposal Practices and Challenges in the United States' *Surface Coal Mining Technology Engineering and Environmental Aspects*, ed Fung R, Noyes Data Corporation, New York, 1981, pp368–374

McCarthy M, 'Quarry firms warned over damage to countryside' *The Times* 3 May 1991

McCarthy RE, 'Surface Mine Siltation Control', *Mining Congress Journal*, June 1973, pp30–35

McGowan G, 'Coping with Environmental Pressure, The Value of Pre-development Planning and Nuisance Control', *Quarry Management*, August 1989, pp33–34

McLoughlin J, 'BACMI winning on aggregate' Construction News, Special Report on Quarrying and Mineral Extraction June 28th 1990 p26

Miles L, 'Waging Coal War', *Surveyor* 25th February 1988 pp12–15

Miller ML et al, 'Mendham Priory, Suffolk, proposed extraction of sand and gravel at Priory Farm', Correspondence 15th September 1989, 5th January 1989 and 7th November 1988 supplied by English Heritage

Miller GG, 'Selective Overburden Placement' *Surface Coal Mining Technology, Engineering and Environmental Aspects*, ed Fung R, Noyes Data Corporation, New York, 1981, pp343–347

Millington G R, 'Environmental Problems Associated with Opencasting' *Clean Air* 11 2, 1981, pp47–54

MJ Carter Associates, 'Low Level Restoration of Sand and Gravel Workings to Agriculture with Permanent Pumping', Report to Department of the Environment HMSO October 1988

Mohanty B et al, 'Optimum Blast Design', *The Northern Miner Magazine* February 1988, pp23–26

Moyes AJ et al, 'Economic comparison of dust collection methods in the quarrying industry', IMM Symposium, Minerals and the Environment Symposium, London, June 1974, pp579–588

Murthy RK et al, 'South Arcot Aquifer and its Utilisation' *Braunkohle* 38 8, August 1986, pp223–226

Muskett CJ, 'Environmental Assessment of Opencast Coal Development', Mine and Quarry Environment 2 3, 1988, pp12–15

Nalpanis P et al, 'Suspension, Transport and deposition of Dust from Stockpiles', Report from Warren Spring for DTI, November 1988, p71

National Rivers Authority, Thames Region, private communications, March 1990, May 1991

National Rivers Authority, South West Region, private communication, March 1990

National Rivers Authority, Welsh Region, private communication, March 1990

National Rivers Authority, Wessex Region, private communication, March 1990

19. References (Alphabetical)

National Rivers Authority, Yorkshire Region, private communication, March 1990

National Federation of Clay Industries Ltd, private communications, August 1990, January & May 1991

National Rivers Authority, Anglian Region, private communication, March 1990

National Rivers Authority, Severn-Trent Region, private communication, March 1990

Nature Conservancy Council, private communications, July 1990 & January 1991

NCB Mining Department, 'Water in the Coal Mining Industry (Technical Management of)' NCB 1982, Ch 18

New BM, 'Ground Vibration Caused by Civil Engineering Works', Research Report 53 of Transport and Road Research Laboratory, Crowthorne, 1986

Nguyen YV et al, 'Identification and Control of Sources of Airborne Coal Dust at Large Storage Coal Piles', paper 80-11.4 73rd Annual Meeting of the Air Pollution Control Association, Montreal June 22-27, 1980

Noise Review Working Party, Department of the Environment 'Report of Noise Review Working Party 1990' HMSO 1990

North R, 'Quarry companies present new face' Independent 23 January 1990

Norton P, 'Future Trends in Mine Design and Restoration in relation to Environmental Constraints', Proceedings of symposium 'Surface Mining -Future Concepts' University of Nottingham 18-20 April 1989, pp124-127

Oates JAH et al, 'The Environmental Effects of Sound from Quarry Blasting', unknown origin

OECD, Environment Committee, Group on the Energy, 'Coal and Environmental Protection (Environmental issues and remedies)', report Paris, 1983

OECD, Environment Committee, Group on the Energy, 'Coal and Environmental Protection (Costs and Costing methods)' report Paris, 1983

Offord C et al, 'Planning and Opencast Coal : Stobswood, Northumberland' Mineral Planning 39 June 1989 pp6-9

Offord C et al, 'Open Cast Coal Mining – A Local Authority Viewpoint', M & Q Environment 3 2, 1989, pp10-14

Olson K S et al 'Fugitive Dust Control for Haulage Roads and Tailing Basins US Bureau of Mines Report of Investigations RI9069, Washington, 1987

Oxford Archaeological Associates Ltd, private communication, January 1990

Oxford Archaeological Unit, private communication, January 1990

Park DG, 'Relocating magnesian limestone grassland' Biological habitat reconstruction, ed Buckley GP, Belhaven Press London 1989

Parkin RJ, 'The Planning, Operational and Environmental Aspects of Opencut Mining at Boundary Hill Mine Central Queensland', The Mining Engineer December 1988, pp268-275

Parrett F, 'The Application of Water-Based Foam in Quarry Dust Suppression', Quarry Management April 1987, pp39-40

Paulson KR, 'The Merger of Engineering and Environmental Technology in Modern Mining and Mineral Processing' Industry and the Environment 8 1, 1985, pp2-4

Peak National Park, private communication, September 1990

Pearce D et al, 'Blueprint for a Green Economy' Report to DoE published by Earthscan Publications Ltd, London, 1989

Pearse G, 'Belt conveyors – increasingly the chosen option' Mining Magazine April 1984 pp362-373.

Peart JD et al, 'Authorisation Procedures for Opencast Mining', Journal Planning & Environmental Law, January 1985, pp11-20

Peltier MA et al, 'Blaster's Training Manual for Metal and Non-metal Miners', Part of Information Circular 9135, 'Surface Mine Blasting', Proceedings: Bureau of Mines Technology Transfer Seminar, Chicago, 1987, p19-24

Philipson G, 'Where do we go from here?', Quarry Management, May 1988, pp43-46

Phillipson R (Interview), 'Industry planning to beat shortfall' Construction News, Special Report on Quarrying and Mineral Extraction, June 28 1990 p23

Pike DC 'Road Track Costs – Big Lorries versus Small' Transport Note TRN2, Sand and Gravel Association Ltd, 1976

Platt JW, 'Environmental control at Avoca Mines Ltd. Ireland' IMM Symposium, Minerals and the Environment, London June 1974, pp731-758

Pollock SHA, 'Coastal Superquarries – from concept to reality. Planning implications of the Glensanda development, Highland Region, Scotland', IMM Symposium, Minescape '88, 1988, pp313-321

Preston CJ, 'Use of Chemical Binders in Dust Control', MSc thesis, Centre for Environmental Technology, Imperial College, 1980

Proctor R 'Putting our energy into nature' British Coal Opencast Executive Publication c1990

Purves JB, 'Aspects of mining and pollution control in SW England', IMM Symposium, Minerals and the Environment, London, June 1974, pp159-179

R & A Young Mining Ltd, 'Proposed Coal Mine at Whitwell' planning application (incorporating Environmental Statement), Application and Accompanying Statement 1, 1990

R & A Young Mining Limited, private communications, January & May 1990

Ralph MO et al, 'Performance of the British Standard Directional Dust Gauge' Warren Spring Laboratory Paper No W89001(PA), Stevenage, January 1989

Rawlinson RD et al, 'Noise from Surface Mineral Workings' Mine & Quarry April 1986, pp26-28

Redbridge LB, private communication, January 1990

Redman M, 'Archaeology and Development', Journal of Planning and Environmental Law January 1990, pp87-98

Rendel S et al, 'Planning Consent for Mineral Extraction and Landfill: The Wildlife Factor' LEB winter 1990/91, pp6-8

Richardson S, 'The Foxhouses Kestrels', Newsletter, Cambria Trust for Nature Conservation, Ambleside, August 88, p3.

19. References (Alphabetical)

Riddleston B, 'Opencast Coal Mining in South Wales – Environmental Issues in Practice' IMM Symposium, Minescape '88, 1988, pp179–189

Ripley EA et al, *Environmental Impact of Mining in Canada*, Centre for Resource Studies Queens University Ontario, 1978

Roberts DI, 'Construction Aggregates and the Cost to Conservation' *Quarry Management and Products*, March 1981 pp167–174.

Robinson MK et al, 'Disposal of Excess Spoil: Durable Rockfill' Proceedings 1983 Symposium on Surface Mining, Hydrology, Sedimentology and Reclamation, Kentucky, November 1983, pp179–188

Rock GA, 'The Environmental Noise Impact of a Large Quarry Extension', Proceedings IOA Meeting 'Noise from drilling, mining and quarrying operations', Bournemouth, April 1989

Rock Environmental Ltd, private communication, July 1991

Rock GA, 'Planning and Noise from Mineral Workings', Proceedings IOA **11** pt 6, 1989, pp27–34

Rosso WA 'Water: Resource or Liability in Surface Mining' Proc Conference & Expo 'Coal Age', Symposium on Surface Coal Mining & Reclamation, Louisville US October 1979, McGraw & Hill, New York 1979, pp99–102

Royal Society for Nature Conservation, 'Losing Ground – Skeletons in the Cupboard – Mineral Planning' RSNC March 1991

Royal Commission on Environmental Pollution, 'Best Practicable Environmental Option' HMSO February 1988

RPS Group PLC 'Planning Application and Environmental Impact Statement for the Extraction of Sand and Gravel from Land North of Hertford' Written Statement and Appendices, non-technical Summary, RPS Group plc, May 1990

Rukavina M, 'Sand and Gravel Scrubbed/Upgraded Without Water', *Rock Products*, October 1989, p21

Rutherford LA 'The Environmental Costs of Opencast Mining: Issues of Liability' *J of Environmental Law*, **2** 2, 1990, pp161–177

Scheiner B J 'New Dewatering Technique for Fine Particle Waste' *XVI International Mineral Processing Congress*, ed E Forssberg, Elsevier Science Publishers BV, Amsterdam 1988, pp1951–1961

Schofield C et al, 'Guide to the handling of Dusty Materials in Ports', British Materials Handling Board, April 1983

Secretary of State for the Environment, decision on 'Coolscar Quarry, Kilnsey, North Yorkshire, Eskett Quarries Limited', M/5069/42/4 10th March 1987

Secretary of State for the Environment, decision on 'Appeal by British Coal Corporation' APP/D1000/A/88/87968 July 1989 relating to Kirk Opencast Site.

Secretary of State for the Environment, decision on 'Appeal by British Coal Corporation APP/N3400/A/86/59987', 6th April 1988

Secretary of State for the Environment, decision on 'Coolscar Quarry, Kilnsey, North Yorkshire' M/5069/42/4 10th March 1987

Secretary of State for the Environment, decision on 'Appeal by British Coal Corporation APP/R2900/A/88/85681', 21st December 1988 related to Linton Colliery.

Secretary of State for the Environment, decision on 'Land at Bridge Farm, Willingham Appeal by Tarmac Roadstone Limited' APP/W0530/A/88/83638, 15 March 1989

Secretary of State for the Environment, decision on 'Appeal by Redland Aggregates Ltd. Application No.N86/577' 12th December 1988

Sheiner B J 'Increasing Dewatering Efficiency through Optmization of the Flocculation Sequence' Proceedings of International Technical Conference on 'Filtration and Separation' Ocean City, US March 1988, ed Frederick E R et al, pp371–378

Shenton V et al, 'The Control of Substances Hazardous to Health – The experience of a major quarrying group', *Quarry Management* October 1989, pp25–30

Shukla J et al, 'Effect of Cement Dust on the Growth and Yield of *Brassica campestris* L.' *Environmental Pollution* **66**, 1990, pp81–88.

Simons DB et al, 'Surface Coal Mining Hydrology', *Minerals and the Environment* **6** 4, December 1984, pp133–144

Siskind DE et al 'Comparative Study of Blasting Vibrations From Indiana Surface Coal Mines' US Bureau of Mines Report of Investigations RI9226, Washington, 1989

Siskind DE et al 'Structure Response and Damage Produced by Ground Vibration From Surface Mine Blasting' US Bureau of Mines Report of Investigations RI 8507, Washington, 1980

Siskind DE et al, 'Noise and Vibrations in Residential Structures From Quarry Production Blasting' US Bureau of Mines Report of Investigations 8168, Washington 1976

Siskind DE et al 'Structural Response and Cosmetic Cracking in Residences from Surface Mine Blasting' SANDIA Report, 59th Shock and Vibration Symposium **1**, October 1988, pp319–334

Siskind DE et al 'Structure Response and Damage Produced by Airblast from Surface Mining' US Bureau of Mines Report of Investigations RI8485, Washington, 1980

Smith BH, 'North Yorkshire CC, Yorkshire Dales National Park, Appeals and Applications by Eskett Quarries Ltd. (Reopened Inquiry)' Report to the Secretary of State for the Environment re Coolscar Quarry, Ref C/783/MB/P 17th August 1985.

Smith M, 'North Pennines Landscape Policy' *Mineral Planning* December 1989, pp25–26

Smith M, 'Taking nature for a ride' *Guardian*, 31 August 1990

Somerset CC 'Ditches and Rhynes – A Guide to Ditch and Rhyne Management for Wildlife' Somerset Levels and Moors, Guidelines No 1, Somerset County Council 1986

Somerset CC, letter 10th October 1990

Somerset CC, 'Environment Committee Proposals for Lorries and the Environment' Somerset CC 1991

Somerville SH, 'Control of Groundwater for Temporary Works', Report 113, CIRIA, London, 1986

South Bucks DC, private communication, February 1990

South Glamorgan CC, County Planning Officer, 'Quarry Monitoring Programme', Report of County Planning Officer to Environmental Planning Committee, 21st May 1979

South Glamorgan CC, letter and extracts from the Council's evidence at a public inquiry into Rhoose Quarry, 9th May 1990

South Glamorgan CC/Mid Glamorgan CC (1982), 'Local Quarry Plan No4. NW Cardiff Radyr/Tongwynlais Area', Joint Consultative Report June 1982

19. References (Alphabetical)

South Glamorgan CC, 'County Structure Plan Studies, Local Quarry Plan No 2, Wenvoe/St Andrews Area', Survey Report Feb 1979 & Draft Local Plan Feb 1979 (amended May 1979, Oct 1982 & April 1985)

South Glamorgan CC, 'Progress to Date 1976-1984', Internal Report South Glamorgan County Council 1984, pp173-204

Stacks JF, 'Stripping – the surface mining of America', a Sierra Club Battlebook, New York, 1972.

Stacks JF 'Social Impact', *Stripping*, Chapter 4, Sierra Club New York 1972, pp51-66

Staffordshire CC, 'Extraction of sand and gravel at Alrewas North, Alrewas, Redlands Aggregates Ltd' report to Planning Committee A.19(L) relating to application of 21.12.89, 1990 pp74-80

Stagg MS et al, 'Effects of Repeated Blasting on a Wood-Frame House', US Bureau of Mines Report of Investigations, RI8896, 1984

Stagg MS et al, 'Influence of Blast Delay Time on Rock Fragmentation: One-Tenth-Scale Tests', Information Circular 9135, 'Surface Mine Blasting', Proceedings: Bureau of Mines Technology Transfer Seminar, Chicago, 1987, pp79-95

Stagg MS et al, 'Effects of Blast Vibration on Construction Material Cracking in Residential Structures', Information Circular 9135, 'Surface Mine Blasting', Proceedings: Bureau of Mines Technology Transfer Seminar, Chicago, 1987, pp32-43

Stanton W, 'Hard Limestone: Too valuable to quarry' *Mineral Planning*, June 1990 pp3-9

Steering Group and Working Group appointed by the Minister of Transport 'Traffic in Towns – A study of the long term problems of traffic in urban areas' HMSO London 1963

Stephenson IC, 'Are All Quarries Unattractive? Not This One...', *Stone Review*, December 1987, pp11-12

Strathclyde RC, private communication, March 1990

Suffolk CC, private communication, November 1989

Surrey CC, private communications, November 1989, January & May 1991

Swain C, 'Mineral Policies in Development Plans' *Mineral Planning* 45 December 1990 pp7-12

Talbot D, 'Recent Advances in Open-pit Blasting, Surface mining experience benefits quarrying', *Quarry Management*, August 1988, pp31-34

The Royal Society for Nature Conservation 'The Peat Report' RSNC c1990

The Royal Society for the Protection of Birds, confidential communication, March 1990

The Opencast Mining Intelligence Group, 'A Reassessment of Opencast Mining' Report for a conference on opencast mining in 1979

The Green Alliance, 'The Environmental Protection Act 1990' Green Alliance, London 1990

The Trust for Wessex Archaeology, 'Lower Farm, Greenham Archaeological Evaluation', Report 1987 published by The Trust for Wessex Archaeology

Thomas JB, 'Developments in Explosives Utilization in ECC', *Quarry Management*, 1986

Thomas W, 'The Landscape Architect and the Coalmining Industry' *The Mining Engineer* February 1989, pp377-381

Thompson J, 'The Compaction of Opencast Backfill, A new option in restoration strategy', *Quarry Management*, January 1990, pp11-16

Timson JP, 'Environmental Aspects of Mine Planning and Development' Annual Review, Irish Association for Economic Geology, 1988, pp56-60

Tollner EW et al, 'The use of grass filters for sediment control in strip mine drainage Volume 1', Report prepared by University of Kentucky Institute of Mining and Mineral Research, 1987

Tomlinson P, 'The Environmental Impact of Opencast Coal Mining' Town *Planning Review*, 53 1, Jan 1982, pp5-28

Tomlinson RV, 'The Public Impact of Quarrying,' *Quarry Management* February 1983, pp85-90

Tompsett KR 'siteNoise – Open site noise prediction system' New Products, *Acoustics Bulletin* July 1990 p40

Trettle C, 'Propagation Law and Small Bore Hole Blasting', *Stone Review* June 1986 pp18-19

Tromans S et al, 'The Environmental Protection Act 1990: Its Relevance to Planning Controls' *J of Planning & Environmental Law* June 1991 pp507-515

Tuck G, 'The control of Airborne Dust in Quarries', *Quarry Management* April 1987, pp 27

Turland MJ, 'Planning Permission Conditions – are they a Waste of Resources?', Correspondence, *Journal of Planning and Environment Law* March 1990, pp184-5

Tyldesley D et al 'Ling Hall Quarry, Lawford Heath Warwickshire – Environmental Impact Statement' for Ideal Aggregates Ltd, March 1990

UK Government 'Town and Country Planning (Assessment of Environmental Effects) Regulations 1988', HMSO 1988

UK Government, 'This Common Inheritance. Britains Environmental Strategy' White Paper, HMSO September 1990

UN ESCAP, 'Proceedings of the working group meeting on environmental management in mineral resource development' Mineral Resource Development Series, 49, 1982

UNEP/IEO, 'Environmental Aspects of Selected Nonferrous Metals Ore Mining', Technical Guide 1st draft, October 1989, Paris

Upton SL et al, 'Some Experiments on Material Dustiness', Paper for presentation at the Aerosol Society Annual Conference Univ Surrey April 1990

van Boven J et al, 'Environmental Compliance at Hoboken in 1983' *Journal of Metals* August 1984, pp62-64

Walker PD, 'Cheshire County Council Application by Dalefords Estates Ltd.' Report to Secretary of State re Moss Farm, Sandiway, Ref A/257X/KP/P December 1985.

Walker AW 'Noise Standards – Where are they?' County Planning Officers Society, Committee no. 3, Conf 'Mineral Planning into the 1990s' Loughborough Univ September 1990

Walker A & Cockcroft P, 'Noise Standards – Where are they?' *Mineral Planning* 46 March 1991, pp3-5

Walker A, 'Opencast Mining Noise', *M & Q Environment* 3 2, 1989, pp20-2

19. References (Alphabetical)

Waller RA 'Planning to Avoid Noise Exposure in Construction' Construction Industry Research & Information Association Technical Note TN 138, 1990

Waller RA, 'A Guide to reducing the exposure of construction workers to noise' CIRIA report 120, London, 1990

Walsall MBC, private communication, July 1990

Walters SG, 'Buckinghamshire CC Minerals Subject Plan – report of a Public Local Inquiry', Report of Inspector 27th August 1982

Ward E 'Levels of Vibration and Air Blast at Warmsworth Quarry' Blasting, *Mineral Planning* **44** September 1990 pp15–18

Wardell Armstrong 'Assessment of the Potential Impacts from Dust Generated during Opencast Mining and Quarrying' Leaflet of Wardell Armstrong Ltd, 24 August 1990

Wells HD, 'An Evaluation of the Sources and Effects of, and Control Mechanisms for, Particulate Emissions ....re Delabole Slate Quarry Cornwall', MSc thesis, Centre for Environmental Technology, Imperial College, London, September 1985

West Glamorgan CC, letter and documents relating to Derlwyn Opencast public inquiry, 22nd May 1990

West J, 'Involve Employees to Build Support – Public Acceptance', *Rock Products*, March 1986, pp54–57

White T, 'Blasting to Specification' *Quarry Management* December 1989, pp27–34

Wiles RA, 'New Publications for Road Systems at Coal Mines' Proceedings 1983 Symposium on Surface Mining, Hydrology, Sedimentology and Reclamation, Kentucky, November 1983, pp283–287

Williams J, 'Fugitive Dust', *Mine and Quarry*, March 1986, pp9–10

Williams J, 'Fresh problems springing up' *Quarrying, Construction News Supplement* June 6 1991

Wilson CI, 'Scottish Superquarries – A catalyst for integrated development?' *Mineral Planning* **47** June 1991 pp18–21

Wilson H, 'A look at current and future trends', *Quarry Management*, April 1988 pp33–35

Wilson K, 'A Guide to the reclamation of mineral workings to Forestry' R & D Paper 141, Forestry Commission 1985

Wilson TJ et al, 'Good Vibrations', *Mine & Quarry*, Jan/Feb 1987, pp51–2

Wilton TJ, 'Air Overpressures from Blasting' Paper published by Rock Environmental Ltd, undated

Wiss JF, 'Construction Vibrations: State-of-the-Art' *Journal of the Geotechnical Engineering Division, Proceedings ASCE* **107** GT2, February 1981, pp167–181

Woodside A et al, 'Control of Dust in Stone Processing – A survey of Northern Ireland plants', *Quarry Management* March 1989, pp29–31

Worthington TR et al 'Transference of Semi-Natural Grassland and Marshland onto Newly Created Landfill' *Biological Conservation* **41** 1987, pp301–311

WS Atkins, 'Noise from Surface Mineral Workings – Community Response Studies: Initial Findings', Draft of work in progress, report to DOE 8/1987 (unpublished)

WS Atkins Engineering Sciences Ltd, 'The Control of Noise at Surface Mineral Workings', Report on behalf of Department of the Environment, HMSO London 1990

WS Atkins Engineering Sciences, 'Draft Notes on Planning Guidance', Second Draft, February 1989 (unpublished)

Zaburunov SA, 'Blasting, Mind over Materials' *Eng & Mining J* April 1990, pp20–25

A computerised database of the literature reviewed during the study has been created. This is available complete with instruction manual at cost price from the Department of the Environment at the address below:

>Department of the Environment,
>Room C16/19a
>2 Marsham Street,
>London SW1P 3EB.